D1160081

Developer's Dilemma

Inside Technology

edited by Wiebe E. Bijker, W. Bernard Carlson, and Trevor Pinch

Casey O'Donnell, *Developer's Dilemma: The Secret World of Videogame Creators*

Christina Dunbar-Hester, *Low Power to the People: Pirates, Protest, and Politics in FM Radio Activism*

Eden Medina, Ivan da Costa Marques, and Christina Holmes, editors, *Beyond Imported Magic: Essays on Science, Technology, and Society in Latin America*

Anique Hommels, Jessica Mesman, and Wiebe E. Bijker, editors, *Vulnerability in Technological Cultures: New Directions in Research and Governance*

Amit Prasad, *Imperial Technoscience: Transnational Histories of MRI in the United States, Britain, and India*

Charis Thompson, *Good Science: The Ethical Choreography of Stem Cell Research*

Tarleton Gillespie, Pablo J. Boczkowski, and Kirsten A. Foot, editors, *Media Technologies: Essays on Communication, Materiality, and Society*

Catelijne Coopmans, Janet Vertesi, Michael Lynch, and Steve Woolgar, editors, *Representation in Scientific Practice Revisited*

Rebecca Slayton, *Arguments that Count: Physics, Computing, and Missile Defense, 1949–2012*

Stathis Arapostathis and Graeme Gooday, *Patently Contestable: Electrical Technologies and Inventor Identities on Trial in Britain*

Jens Lachmund, *Greening Berlin: The Co- Production of Science, Politics, and Urban Nature*

Chikako Takeshita, *The Global Biopolitics of the IUD: How Science Constructs Contraceptive Users and Women's Bodies*

Cyrus C. M. Mody, *Instrumental Community: Probe Microscopy and the Path to Nanotechnology*

Morana Alač, *Handling Digital Brains: A Laboratory Study of Multimodal Semiotic Interaction in the Age of Computers*

Gabrielle Hecht, editor, *Entangled Geographies: Empire and Technopolitics in the Global Cold War*

Michael E. Gorman, editor, *Trading Zones and Interactional Expertise: Creating New Kinds of Collaboration*

Matthias Gross, *Ignorance and Surprise: Science, Society, and Ecological Design*

Andrew Feenberg, *Between Reason and Experience: Essays in Technology and Modernity*

Wiebe E. Bijker, Roland Bal, and Ruud Hendricks, *The Paradox of Scientific Authority: The Role of Scientific Advice in Democracies*

Park Doing, *Velvet Revolution at the Synchrotron: Biology, Physics, and Change in Science*

Gabrielle Hecht, *The Radiance of France: Nuclear Power and National Identity after World War II*

Richard Rottenburg, *Far-Fetched Facts: A Parable of Development Aid*

Michel Callon, Pierre Lascoumes, and Yannick Barthe, *Acting in an Uncertain World: An Essay on Technical Democracy*

Ruth Oldenziel and Karin Zachmann, editors, *Cold War Kitchen: Americanization, Technology, and European Users*

Deborah G. Johnson and Jameson W. Wetmore, editors, *Technology and Society: Building Our Sociotechnical Future*

Trevor Pinch and Richard Swedberg, editors, *Living in a Material World: Economic Sociology Meets Science and Technology Studies*

Christopher R. Henke, *Cultivating Science, Harvesting Power: Science and Industrial Agriculture in California*

Helga Nowotny, *Insatiable Curiosity: Innovation in a Fragile Future*

Karin Bijsterveld, *Mechanical Sound: Technology, Culture, and Public Problems of Noise in the Twentieth Century*

Peter D. Norton, *Fighting Traffic: The Dawn of the Motor Age in the American City*

Joshua M. Greenberg,
From Betamax to Blockbuster: Video Stores tand the Invention of Movies on Video

Mikael Hård and Thomas J. Misa, editors, *Urban Machinery: Inside Modern European Cities*

Christine Hine, *Systematics as Cyberscience: Computers, Change, and Continuity in Science*

Wesley Shrum, Joel Genuth, and Ivan Chompalov, *Structures of Scientific Collaboration*

Shobita Parthasarathy, *Building Genetic Medicine: Breast Cancer, Technology, and the Comparative Politics of Health Care*

Kristen Haring, *Ham Radio's Technical Culture*

Atsushi Akera, *Calculating a Natural World: Scientists, Engineers and Computers during the Rise of U.S. Cold War Research*

Donald MacKenzie, *An Engine, Not a Camera: How Financial Models Shape Markets*

Geoffrey C. Bowker, *Memory Practices in the Sciences*

Christophe Lécuyer, *Making Silicon Valley: Innovation and the Growth of High Tech, 1930–1970*

Anique Hommels, *Unbuilding Cities: Obduracy in Urban Sociotechnical Change*

David Kaiser, editor, *Pedagogy and the Practice of Science: Historical and Contemporary Perspectives*

Charis Thompson, *Making Parents: The Ontological Choreography of Reproductive Technology*

Pablo J. Boczkowski, *Digitizing the News: Innovation in Online Newspapers*

Dominique Vinck, editor, *Everyday Engineering: An Ethnography of Design and Innovation*

Nelly Oudshoorn and Trevor Pinch, editors, *How Users Matter: The Co-Construction of Users and Technology*

Peter Keating and Alberto Cambrosio, *Biomedical Platforms: Realigning the Normal and the Pathological in Late-Twentieth-Century Medicine*

Paul Rosen, *Framing Production: Technology, Culture, and Change in the British Bicycle Industry*

Maggie Mort, *Building the Trident Network: A Study of the Enrollment of People, Knowledge, and Machines*

Donald MacKenzie, *Mechanizing Proof: Computing, Risk, and Trust*

Geoffrey C. Bowker and Susan Leigh Star, *Sorting Things Out: Classification and Its Consequences*

Charles Bazerman, *The Languages of Edison's Light*

Janet Abbate, *Inventing the Internet*

Herbert Gottweis, *Governing Molecules: The Discursive Politics of Genetic Engineering in Europe and the United States*

Kathryn Henderson, *On Line and On Paper: Visual Representation, Visual Culture, and Computer Graphics in Design Engineering*

Susanne K. Schmidt and Raymund Werle, *Coordinating Technology: Studies in the International Standardization of Telecommunications*

Marc Berg, *Rationalizing Medical Work: Decision Support Techniques and Medical Practices*

Eda Kranakis, *Constructing a Bridge: An Exploration of Engineering Culture, Design, and Research in Nineteenth-Century France and America*

Paul N. Edwards, *The Closed World: Computers and the Politics of Discourse in Cold War America*

Donald MacKenzie, *Knowing Machines: Essays on Technical Change*

Wiebe E. Bijker, *Of Bicycles, Bakelites, and Bulbs: Toward a Theory of Sociotechnical Change*

Louis L. Bucciarelli, *Designing Engineers*

Geoffrey C. Bowker, *Science on the Run: Information Management and Industrial Geophysics at Schlumberger, 1920-1940*

Wiebe E. Bijker and John Law, editors, *Shaping Technology / Building Society: Studies in Sociotechnical Change*

Stuart Blume, *Insight and Industry: On the Dynamics of Technological Change in Medicine*

Donald MacKenzie, *Inventing Accuracy: A Historical Sociology of Nuclear Missile Guidance*

Pamela E. Mack, *Viewing the Earth: The Social Construction of the Landsat Satellite System*

H. M. Collins, *Artificial Experts: Social Knowledge and Intelligent Machines*

http://mitpress.mit.edu/books/series/inside-technology

Developer's Dilemma

The Secret World of Videogame Creators

Casey O'Donnell

The MIT Press
Cambridge, Massachusetts
London, England

This book was set in Stone Sans Std and Stone Serif Std by Toppan Best-set Premedia Limited, Hong Kong. Printed and bound in the United States of America.

Library of Congress Cataloging-in-Publication Data

O'Donnell, Casey, 1979–
Developer's dilemma : the secret world of videogame creators / Casey O'Donnell.
pages cm.—(Inside technology)
Includes bibliographical references and index.
ISBN 978-0-262-02819-6 (hardcover : alk. paper) 1. Computer games—Programming. 2. Computer software—Development. 3. Video games–Design. 4. Video games—Authorship. 5. Computer software developers. I. Title.
QA76.76.C672O36 2014
794.8'1526—dc23
2014013210

Contents

How to Play (Use) This Game (Book)

The structure of this book is performative. It imitates the level structure of the Nintendo Entertainment System game *Super Mario Bros.* (*SMB*) to highlight the importance that games and how an understanding of games does work in game developer culture. This sense of a shared history and experience provides foundations for how videogame developers talk about their occupations. This is not really any different from any other discipline or environment of cultural production where experiences and language become entangled in ways that prevent broader accessibility. To make the work experiences of developers decipherable, the book is structured in a way that provides readers with the tools to help debug game developer culture.

WORLD X—Each World in this book is a "chapter" in the traditional sense.

WORLD X-1–X-4—Each World is a collection of four levels, much like *SMB*. Each level advances material that culminates in the Boss Fights found at the end of each world.

BOSS FIGHT—Boss Fights, in games, require that players draw on new skills, lessons, or mechanics in order to progress to the next World. Think of it like a test. Bonus: I get to take the academic gloves off.

#: SET DEBUG_MODE = 1—Most games have, lying underneath, a host of tools/options/data used during development that is invisible to the player. By turning "DEBUG_MODE" on (to 1 or TRUE), I am offering up some of the empirical data that lies behind the text.

AUTHOR_DEV_DILEMMA—I use these sections, contained only in DEBUG_MODEs, to delineate my reflective comments. It is interpretation that came after the gathering of the data being presented. Time, experience, and analytic perspective offer something absent in just the transcripts.

CASEY—My voice in interview transcripts contained only in DEBUG_MODEs.

TITLE_Project—Informant's words in DEBUG_MODEs appear in a fashion that indicates their relative position within a company and the name or codename of the project they were working on at the time.

#: SET DEMO Switch_MODE = 0—Turn DEMO Switch_MODE off (to 0 or FALSE)

Minus World: A Glitch

World –1: "Ship It"

There exists a glitch in the original *Super Mario Bros.* game for the Nintendo Entertainment System (NES). If the player performs a series of actions in World 1-2 of the game (the proper sequence of events is left as an exercise to interested readers) they enter World –1 ("Minus World"). The catch, however, is that once inserted into "Minus World," they cannot escape. The player is ultimately doomed. They consistently die due to either the creatures in the level or the continuous countdown of the clock in the game, which doesn't reset at the end of a level, as in the typical game. Worse, the levels loop together, forever returning the player to the start. There is no escape from Minus World.

Minus World is a bit like trying to write a book about the game industry and the people that work in it. No matter how hard an analyst tries to capture it all, things continue to move and shift. Yet, that's precisely what makes this text an important contribution to the historical memory of the game industry, a memory that I argue in this text is under-documented, rarely studied, and widely misunderstood. I could have returned to the material in this text year after year, modifying and adjusting according to the shifts of independent and industry game development each year, much like Mario or Luigi continually returning to the looping worlds of Minus World. New consoles, tools, and platforms will surely render some of the ethnographic material in this text "quaint" over time, yet without this text, the idea that such an account is important would never even be considered. For this reason, I must eventually simply say, "Ship it."

Visibility of developers has changed significantly in recent years, thanks in great deal to the micro-blogging site Twitter. Game developers can be found and engaged with in ways that were largely impossible prior to 2007. Twitter use by developers is one of several subsequent game-developer-focused studies now under way. I am certain that the site will play a role in the future of this text, precisely because of the networks that developers have solidified in that space. If I thought having my informants "talk back" during my fieldwork was complicated, I can only imagine the kinds of comments that may find their way to me [@caseyodonnell] in 140 character morsels. Game developers increasingly blog on their own or on sites like Gamasutra, which occasionally selects entries from developers that use the site to "feature" insights from numerous perspectives across the industry. This largely was not the case when I began this endeavor.

Also captured in the trope of Minus World are the glitches found in any massive creative endeavor, like this text. There are sure to be errors or bugs to be found. Those, along with the analysis and conclusions reached, are subject to my particular position and context. As one of a very small number of cultural analysts that have negotiated access to game developer communities, my perspective is limited and finite.

Lest that caveat sound apologetic, I must reassert my firm belief that all critiques and Boss Fights in the following text result from the sincere hope of bringing productive change to the work of game development and the game industry more broadly. I want to make the creative collaborative work of my informants more visible because what every game developer does every day can inform so many others. Making game development visible, rather than a secret diary kept quietly locked away, will do more to address quality of life struggles. Writing with this aim foremost in my mind, I try to honor my informants. Their work is indicative of what labor has become in our current historical and cultural moment. I thank the game development community for the opportunity to perform my research, and I hope that the resulting product contributes something positive. Without a historical memory, I worry about the trajectory that such a headlong rush forward will result in, particularly at a moment when independent games, serious games, art games, games for impact, and numerous others are finally finding a foothold.

My game industry informants from across the world made this text possible. Especially important are those who made access to their sites possible. I would like to thank Vicarious Visions, 1st Playable, Dhruva Interactive, RedOctane India (now closed) and all the other sites and developers who spoke with me during my research. Such an undertaking would have been unfeasible without their candid participation in this project. To the handful of people who helped me navigate initial access, a special thanks is necessary. Even more so to those of you that have continued the conversation in subsequent years. You know who you are. You are awesome. My various field sites' willingness to provide access, resources, and copious amounts of coffee made this research possible and without such generosity this book would not exist. Their contribution is a testament to precisely the kind of creative collaborative interdisciplinary work that lies at the core of this book. Thank you for playing the *Developer's Dilemma* well.

I owe a debt of gratitude to the faculty and graduate students of Rensselaer Polytechnic Institute's Science and Technology Studies Department. Most important, I am thankful for the work of my dissertation committee. Each member, in his or her own way, influenced the predecessor to this text (O'Donnell 2008). Kim Fortun, Mike Fortun, Nancy Campbell, Atsushi Akera, and Christopher Kelty: thank you. Special thanks goes to my chair, Kim, who helped me understand the underlying game mechanics of academia. Kim, Mike, Kora, and Lena, you were our adopted family while living in Troy, and I thank you for that.

My editor, Christine Harkin (http://christineharkin.com) helped propel this book out of stasis in the winter and spring of 2012, for which I am eternally grateful. I am appreciative for the hard work of the anonymous reviewers that looked at the early versions of the text, even when it remained a train wreck. Much like how a game under development can appear to its developers as a "busted pile of crap" (Wyatt 2012), such was the way I viewed this text until Christine and I were able to coax from the wreckage what now stands as *Developer's Dilemma*. Thanks also to Marguerite Avery and the MIT Press for their continued interest in this text, even while it was under revision.

To my mom, thank you for all the random conversations about poststructuralism at odd hours and for knowing me better than sometimes I know myself. To my mother-in-law, thank you for being the ever-enthusiastic supporter, helper on the home front, and for keeping the reality of day-to-day corporate culture in my frame.

I think most authors have an animal companion that deserves recognition. I have had the luxury of three Siamese cats, each has offered input on this book, mostly in the form of hair on my lap. Rambo and Scoshi kept me company through the first two drafts of this text and Ash bit the heels of hands through the completion of the third.

Utmost thanks goes to my wife, Andrea, who has had to live with me throughout the various iterations of this text. She has been with me through the ups and downs of industry life and allowed herself to be dragged across the United States, all the while building her own career. She has helped me find games in the most unlikely places and reminds me of the value of being a player, a coach, and referee, but never all of them at one time. She makes me a better person, scholar, husband, and dad. To Alexis and Caelyn, whom I hope to "classically train," with the gift of an Atari 2600 at age three, a Nintendo Entertainment System at age four, a Sega Genesis at age five, and so on, I'm sorry in advance if all of that results in only future conversations with a psychiatrist. I love you both.

Ship it.

Introduction: A Videogame Industry Primer

World 1: A Tutorial Level

Box 1.1

#: SET DEMO_MODE 1
AUTHOR_DEV_DILEMMA: The game industry lends itself to high-stakes dramas: long hours, looming deadlines, hardcore engineers, big money payouts, and tremendous losses. Yet, rarely have ethnographers or social analysts gained significant and sustained access to these companies. Game development companies don't even lend themselves to inquiry, with significant limitations and non-disclosures that make research problematic. This text is a result of extensive ethnographic fieldwork among game developers working in "AAA" and independent game studios in the United States and India. Though my research as a videogame development ethnographer continues, this text is largely based on material gathered from 2004 to 2008. While the study of games has exploded in recent years, game developers and game development has remained less explored.

This text is due in part to the serendipity of how and why a former engineer turned anthropologist of science and technology began studying game development as a way to ask questions about the new economy and new media work. I was an engineer who had worked in the industry, yet I'd created a path towards social analyst. For me it was about creating cat's cradles (more on that in a bit) with my informants such that I could contribute back in meaningful ways. While I had ostensibly left the game industry for academia, I instead found myself in a position capable of offering constructive criticism able to speak about the structures that game developers experience every day. Those structures are worthy of critique when they rob game developers of happiness and fulfillment in a workplace they are so passionate about.

World 1 marks the entry point into the activities of videogame developers. Much like a player just starting a game, readers need

(*Continued*)

```
a tutorial or introduction because without a primer, the player/
reader is left with no idea of what they can or are expected to
do/learn. World 1 is this text's tutorial.
#: SET DEMO_MODE 0
```

World 1-1: The Text's Software Development Kit (SDK)

Developer's Dilemma makes three major contributions to the field of science and technology studies: empirical, theoretical/conceptual, and methodological. The first contribution lies in the breadth and depth of my research. Scholars have not yet engaged in empirical studies of videogame development practice, and in-depth ethnographic fieldwork in the videogame industry has been absent. Some have looked glancingly at the work of media industries (the game industry included) but none have taken an in-depth ethnographic look at the collaborative work and tools of the videogame industry. This text closely examines the social and technical milieu of videogame developers and where those elements intersect, more broadly, with the numerous other systems and institutions that seem to imbricate "the industry."

The second contribution of *Developer's Dilemma* is a demonstration of how the creative collaborative practice of game developers and game development work sheds new conceptual light on our understandings of work, the organization of work, and the market forces that shape and are shaped by media industries in the new economy. Videogame developers—programmers, artists, game designers, and managers—and videogame development in the United States and India are used as windows into understanding these complex issues.

This text is based on foundational ethnographic data formed from four years of participant observation at game studios in the United States and several months of fieldwork with game studios in India. This is further supplemented by more than seventy in-depth interviews and internal documentation, practices, and protocols from each field site. Patent documents, legal cases, U.S. Securities and Exchange Commission (SEC) filings, and press releases serve to further illuminate the forces and activities of game developers. Trade press and "enthusiast" press material is also used as a means to validate and further contextualize the ethnographic data.[1] This third contribution, largely methodological, grapples with research located in corporate field sites often encumbered by numerous limitations, such as non-disclosure agreements (NDAs).

New media work, exemplified by game development practice, is dependent upon producing new modes of creative collaborative work practice. This book connects the diverse forces and activities (e.g., laws, technologies, and workplace cultures) that make creative collaborative practice central to the way the new economy works.

The way these practices play out and the structural conditions they unfold within, however, simultaneously undercut creative collaborative practice. The core of creative collaborative practice is the ability and necessity of being able to play with and get at underlying systems—technical, conceptual, and social. When access to underlying systems is undermined, so, too, is creative collaborative practice. By making collaborative practice the central concern, this project demonstrates how diverse systems across multiple scales come together in the context of new media work in ways that either help or hurt the function of creative collaborative practice.

Broadly considered, this text draws on three primary bodies of literature: studies of work practice and science and technology production; media studies and game studies; and studies of globalization and the new economy. I draw connections between the relevant literature throughout the text and my empirical material. To provide scaffolding, however, I offer an introductory contextualization below. This initial position among the primary bodies of scholarship being drawn upon is done to provide academic readers with a framework for approaching this text.

Box 1.2

```
#: SET WARNING_LEVEL 10
//-----------------------------------------------------------
// For those of you here just to read about game development,
// do not let the next few paragraphs deter you from the text.
// The more readable section will continue shortly. The
// conceptual parts of the text are there to help situate my
// analysis of the material. I'll try to keep things readable.
//-----------------------------------------------------------
#: SET WARNING_LEVEL 0
```

The primary assemblage of literature situating this text includes the studies of work and work practice. In this sphere are studies of technological and scientific production, since they provide a wealth of theoretical and methodological resources for making sense of the activity of videogame development (Latour and Woolgar 1986; Pickering 1995; Forsythe

2001). These texts demonstrate the often-neglected social and technological aspects that disappear behind completed science and technology. They demonstrate the contingent and constructed character of these endeavors. Texts included in this category also point to the influence of political and economic aspects on the lives and approaches used by practitioners. The way gender and social networks structure labs and workplaces offers significant resources for examining communities of game developers. Further supplementing this category are studies of work and work practice (Suchman 1995; Orr 1991; Barley and Orr 1997a). These texts establish both the importance of studying the everyday lives of working people, as well as studying what work has become in recent times. In many cases, work is much more complex and nuanced than acknowledged by management or those external to those professions. What is typically portrayed as simple or straightforward is often quite the opposite. Even "obvious" problems require much more skill and ability than is recognized.

Anthropological and sociological inquiries into work and work organization offer a wealth of theoretical and empirical resources from which to draw on. In particular, the examination of technical, engineering, and (new) media work indicate a significant disconnect between how work is imagined by those laboring in them. These "new breeds" of workers "violate our concepts for making social sense of work" (Barley 1996, 412). It is with this in mind that researchers have attempted to better understand the relationship workers have with their work, wondering, "what are we to make of someone who says they love their work and cannot imagine doing anything they enjoy more, yet earns so little that they can never take a holiday, let alone afford insurance or a pension" (Gill 2007, 9). While it might seem at first glance that these jobs are different, "hot," "cool," or unpredictable (Neff et al. 2005) in ways that make them less like work and more like play, these types of jobs, work, and organizations are often cited as exemplars of our "Brave New World of Work" (Castells 1998; Beck 2000).

Rather than signifying these workplaces as distinct or different, all indications seem to be that this form of work and the organizations that support them "may become the modal form of work for the twenty-first century" (Barley and Orr 1997b). So while these forms of work seem to be dramatically important in the context of the new economy, they also prove significantly problematic for existing forms of organization. Initially some cultural analysts assumed that a natural transition toward "horizontally" organized work would result in more horizontal forms of organization (Whalley and Barley 1997). Rather, it seems that these changes have proven more contested; that despite "horizontal work processes, collaboration, rather than command, is the key to getting work done" (Zabusky 1997, 130) most organizations balk at the necessary autonomy and trust that must be placed in the individual (Barley 1996). Some organizations attempt a kind of "industrialization of bohemia" that while at a surface plugs into the self-image of workers interested in these emerging industries actually tends to be quite detrimental to workers lives outside the workplace (Ross 2003).

What many of these investigations agree upon is that more research in the corporate context are needed, precisely when they are becoming more and more difficult to perform (Smith 2001). More needs to be learned about how work gets done in contexts where the work of individuals frequently becomes lost or invisible, and in these new horizontal technology organizations such cloaking occurs frequently (Downey 2001). Many researchers have attempted to reconstruct work practice in ways that encourage greater attention to the collaborative social aspects of the workplace (Suchman 1995; Suchman et al. 1999). These fieldwork-centered inquiries indicate that this collaborative and social aspect of work in the new economy makes simplistic approaches to globalization, offshoring, and management particularly difficult (Hakken 2000b). Along this same vein, this research attempts to resocialize and use the creative collaborative efforts of game developers as a means to rethink work in the new economy.

Tied closely to anthropological studies of work are ethnographic studies of organizations and industries closely related to the videogame industry. Because of their intersection with technoscientific practice, many of these projects are highly informed by the broader field of STS. Foundational anthropological studies of engineers (Downey 1998), Linden Labs (Malaby 2009) and the Gaming Industry (Schüll 2012) further inform and situate this study. Each differs in important ways from this study, yet each offers a methodological and conceptual core for "critical participation" (Downey 1998, xi) that many studies like this one attempt to manage. Linden Labs was, in many respects, a game engine developer, though they understood themselves as virtual world architect and host. Rather, they were in the business of creating tools for users to generate content in a virtual world they managed. Engineering culture certainly intersects with the work of game development, but it is precisely the intersection of engineering worlds with art, design and business perspectives that makes the study of game development distinct (O'Donnell 2012). The gaming (as in gambling machines) industry has certainly paid attention to the work of game developers and game scholars, but to different ends, both technologically and culturally. Each serves as an important connection for this study.

Media studies and the emerging discipline of game studies is the second body of scholarship that situates this text. Though this study does not explicitly engage with the images or games produced by game developers, the insight and research in these areas provide a wealth of resources. Studies of online spaces and the gamers that inhabit them provide some insight into the perspective of developers, because most are avid gamers. They provide a foundation for understanding how or why a developer might make some decisions or pursue particular interests over others. Some studies look explicitly at the politics and economics of online worlds created by developers (Castronova 2005; Nardi 2010; Taylor 2006a) and others examine the different ways people play online or offline and how aspects like gender affect these (Cassell and Jenkins 2000). Some have made the turn more explicitly toward production and how user created game modifications (mods) shape play spaces (Taylor 2006b).

Some texts examine the issue of play and games, which is central to the theoretical foundation of this text. How and why people play, or the human or animal propensity for games and play, offer extensive resources from which to draw (Huizinga 1971; Sutton-Smith 1998; Burghardt 2005). In this book, these texts are put into conversation with post-Marxist and cultural studies conceptual frameworks that provide new resources for understanding hegemonic structures and hegemonic projects (Omi and Winant 1994; Hall 1996). Finally, a handful of research projects actually come into contact with game developers, though primarily in small doses, corporate approved doses, or based on fictional situations (Chaplin and Ruby 2005; Coupland 2006; Wark 2007). They point to some of the ambiguity and difficulty of working in the game industry, but stop short of offering new empirical perspectives. Most never make it beyond the big names and highly publicized meltdowns.

A small, emerging set of literature attempts to examine the worlds of videogame developers specifically. These texts are limited by their small sample sizes and limited access to field sites. The majority relies on limited access to online forums or other virtual means that result in a very engineering-centric picture of what videogame development looks like (McAllister 2004). This would make sense given that (as is noted in this text), many aspiring game developers end up pursuing engineering-related resources though the disciplines found in the game industry are actually quite varied. Available web resources reflect the external bias and so does the analysis of those studies. Other studies attempt to be too encompassing, ultimately offering an un-nuanced, general account of a strikingly dynamic industry (Deuze 2007). Other texts work to capture the massive scale and global character of the game industry, but in so doing neglect the daily work activity of game developers (Dyer-Witheford 1999; Dyer-Witheford and Sharman 2005; Dyer-Witheford and de Peuter 2009). Ultimately, the failing in game studies has been close attention to what I call simply "production" or the work involved in creating videogames. The focus has been largely on game experience, game design, game economies, and the feedback loop between player and game. Stepping in closer, I ask a very common STS question: What about the people that create the thing you are studying?

This text draws on a final set of literature from studies and accounts of the new economy and globalization. These texts vary greatly in their empirical engagement with the new economy and, as such, it is most useful to explicate their connection within this text. Some texts focus on modernity, postmodernity, and the changing position of the state (Lyotard 1984; Harvey 1990; Appadurai 1996). These texts often include insight into the new economy, as it is one aspect of, or perhaps a result of (post)modernity ushered in by new communications technologies. Some point to the decline of the state and resulting consequences, like the rise of neoliberalism and the commercialization of formerly government-run institutions. Others look at the new means and mechanisms by which organizations discipline workers and one another or make use of global differentials in monetary systems.

Several texts look explicitly at the process of the information or new economy and globalization (Tsing 2005; Kelly 2006; Varma 2006). These texts exemplify the importance of understanding new global processes, and provide readers with an appreciation for the complexity of the new economy by examining the interweaving of corporate interests in historical and social processes, the global movement and training of new generations of workers for the new economy, the different ways in which global workers are viewed and encouraged or discouraged from working together, and how globalization is experienced on the ground where conflict is experienced.

While grounded in these different approaches to work practice, game theory, and new economy studies, this document differs from most other media that has covered the videogame industry. A small number of publications and online websites cater to game developers, offering new methods or reviews of development tools. Occasionally magazines like *Wired*, *Newsweek*, or *Time* will engage with the game industry, but infrequently with game studios. Journalists will swarm the most well-known executives or game designers, but never rank-and-file developers. Entire magazines are devoted to the latest videogames in development or recently released, and perhaps interview the games producer. The online enthusiast press observes all the meanderings of videogame corporations but offers very little analysis. Each of these perspectives is useful, and can frequently access information and people that I cannot. In this respect they have been invaluable resources.

However, this text is different in that the focal point remains on "typical" developers and work practice—the people who devote the majority of their time to bringing videogames into reality. This text is also executed with an eye to better understanding why they work in the ways they do. It is about observing the activities of everyday developers, who have largely disappeared behind the names of publishers or console manufacturers, and to better comprehending why things go right or wrong.

World 1-2: The Characters—My Gorillas

I never had any intention of studying the game industry. I just wanted to study how software development unfolds in practice. Then I stepped into the offices of Vicarious Visions (VV) in September 2004. For the most part, I was open to studying any software company. I had come expecting to study the lived reality of work practice in the context of the so-called new economy. And for my purposes, VV was ideal: a medium-sized, independent game development studio employing roughly seventy-five employees, a mixture largely of artists, engineers, designers, various managers, and support staff.

Throughout my fieldwork at VV, my informants did not know how to define my position. The inability to place me within existing understandings of what and who counted as legitimate members of the videogame development community was problematic for many of my informants. Often, this manifested in humorous ways, discussions of tribes, gorillas (silverbacks in particular), pith helmets, and mating rituals were common, representing their assumptions about cultural anthropology. Thinking of them this way is not intended to be derogatory, but is a means of expressing how the developers I spent so much time with came to understand the place of an anthropologist and fieldworker among them. It is also an expression of a continual kind of self-jesting that occurs among game developers. A tools engineer from VV actually presented me with a pith helmet at the conclusion of my dissertation defense, an object that now occupies a special place in my office. It, too, is indicative of the kind of creative and satirical humor entrenched in the game industry.

Other developers in the company feared I represented someone determining just how much time they were wasting or whether they were expendable. For such people I was a threat and was kept at arm's length. A few simply could not comprehend what value might be found in observing their communities. Despite my best attempts to explain, they felt speaking with me would simply not be useful. To the remaining people, frequently those who became key informants, I represented a break or schism in a system they felt they could not critique. Somehow I had been authorized to ask the questions that they could not. I represented an opportunity to explicitly reconnect work experience to the political economy within which game development is nestled.

The timing of my arrival at VV was just right, because at the time, development had just begun on a game for Sony's as-yet-unreleased handheld console videogaming system, PlayStation Portable (PSP). The game was based on a forthcoming movie from a major movie studio in partnership with one of the largest comic book companies (among other companies) and VV had been contracted for the project by one of the larger game publishing companies. It was presumed that the title would be developed and released simultaneously with the June 2005 release of the movie and shortly after the March release of the new game system. The project, code-named "*Asylum*," was tasked with producing a series of prototypes and levels that would then be reviewed to determine if the rest of the development work would be entrusted to VV. Many small game studios that have proven their ability to bring a game to market will take on jobs like these either as the mainstay of their work or to fund other internal projects.

Asylum was behind schedule even before there *was* a schedule. Another development studio had long since begun work creating a version of the game for Microsoft's Xbox, Nintendo's GameCube, and Sony's PlayStation 2 game consoles.[2] The PSP and its software was an afterthought, a last-minute realization that a possible market for consumption might be missed. I watched for four months as a team of talented engineers, artists, and designers toiled to create a prototype suitable to justify the remainder of the project. VV's employees each had their own reasons for their enthusiastic work on *Asylum*. For some of the VV team it was a chance to play with hardware that was only available to a handful of game developers. For others it was a chance to work on a "real" console title rather than on systems with greater hardware limitations. Others were excited about the opportunity to work on a game title linked to a blockbuster movie. And some developers felt a passion for the comic book characters contained in the title. For all these reasons, the team toiled day and night for four months creating the foundations for a new game for a new game system.

It was this coalescing of corporate interests and developers' desires that caught my attention. If anything could be said to characterize new modes of work/play, it was precisely this sort of interplay. Daily practices shrouded in secrecy lent an air of mystique to game developers' worlds. Yet, those same practices exhibited a special propensity for collapse under the extreme time pressures that descended from intellectual property (IP) holders and other parties with stakes in the action.[3]

The atmosphere at the office had grown tense by December 2004. *Asylum*'s preproduction had come a long way, but there were many moving parts that had to behave well for the game elements to come together. Code from engineers had to be in place to display special effects overlays created by artists. Animations from artists could not be displayed until the requisite data from designers were added to configuration files. Engineers were waiting for software development kit (SDK) updates from Sony to fix bugs found in the hardware or firmware of the PSP. By this time in the project, frustrated outbursts were common, with one element of the game breaking as other components were added. The automated build and "smoke test" system frequently seemed to be broken and everyone was constantly in hurry-up-and-wait mode. A couple of beers late at night while a daily build was executing sometimes took the edge off, though only for a few short hours before the process began again the next day. In late December 2004 a build was delivered to the publisher for evaluation. The numerous complex software systems, which were expected to be real-time and interactive, strained under the pressure of technologies in transition and the demands of developers. The experimental practices and instrumentality of work/play was quickly transitioning into the realm of "crunch" or mandatory overtime.

In January 2005, five months after I had begun my pilot research, Activision, Inc. (ATVI), one of the largest videogame publishers and a direct competitor of the publisher that had contracted the development of *Asylum*, bought VV. Most of my informants learned of the acquisition just hours prior to the press release; I was informed by the release itself.

I asked the lead designer on *Asylum*, who had aided me in getting access to VV, "What's happening with *Asylum* and what's happening with all of you?" The answers came back within hours: the acquisition by ATVI and the looming release of the PSP had convinced someone somewhere that *Asylum* would likely be unsuccessful, and the project had been shut down. ATVI was assigning everyone to a new project.

It was at this moment that I realized simply talking about "work practice in the new economy" was a woefully inadequate way to look at the project before me. This was the videogame industry, no matter how uncomfortable I was labeling the full scope of my research. Many of my emerging core categories and distinguishing characteristics—such as work/play, interactivity, inter/intranetworks, experimentation, and collaborative practice—were being further complicated through corporate acquisition and consolidation.

In April 2005, Farrar, Straus and Giroux released the book *The World Is Flat: A Brief History of the Twenty-First Century* by Thomas Friedman just as game development work was spreading to other parts of the globe and the two brothers who started VV traveled to China and India to speak with aspiring game development entrepreneurs. It was not lost on my informants or on me that offshore outsourcing loomed just outside the frame.

To better understand this "flat" global new economy I decided to travel to one of these emerging game development sites. I chose India over China, Korea, and Vietnam because I would have more access to English-speaking informants there. Also crucial was an introduction by the Studio Head of VV to the Studio Head of Dhruva Interactive, the self-titled "premier game company in India," situated in Bangalore. The rapid growth of the IT sector in India, India's exploding outsourcing developments, the United States' dramatization of Indian workers in popular media[4] and Dhruva's willingness to grant me site access clarified my choice.

It took eighteen months to secure funding, access, and time to travel to India. During that time I was able to find some Indian development studios willing to provide me site access and others willing only to speak with me upon my arrival. I also accepted an offer to perform fieldwork at RedOctane India in Chennai.

In November 2006, with the assistance of a National Science Foundation (NSF) Grant,[5] I traveled to India to better understand what game development looked like in a seemingly surging new home for game development. For two months I scrambled from field site to field site speaking with employees and studio heads of game development companies, and with aspiring developers in Bangalore, Hyderabad, and Chennai. I performed more than two hundred hours of participant observation, ran more than twenty structured interviews, and had after-work conversations with every game developer willing to share their perspectives of the global game industry that I could find. Two studios, Dhruva and RedOctane, allowed me on-site for one month each. FXLabs in Hyderabad, GameLoft Hyderabad, and

Microsoft's Casual Games Group in Hyderabad were also accommodating, though reluctant to allow me to perform fieldwork over an extended period of time.

In addition to formal observation, I spent my free moments conversing with more individuals interested in working in this emerging area of India's booming IT sector. Many simply wanted to know more about US development practices. They lamented the small number of resources available that might help them make games at a level beyond that of a hobbyist, for though they could find snippets of information here and there, they could use very little of it to inform themselves and their projects. I shared my Firefox game development bookmarks and news feeds with more people than I can recall.

My time with Indian developers taught me that US and Indian developers often face similar issues, though the latter's challenges are frequently exacerbated by temporal and physical distance from the numerous networks and secrets that structure the industry. In some cases this distance is a positive barrier in that it protects Indian developers from the secret society syndrome rampant among US developers, an issue I explore throughout this text.

The biggest commonality among international developers is a shared interest in learning more about what game developers do and how they can each get better at doing it. Despite this shared experience, however, I often found myself wondering, both in India and the United States, "How could anyone assert that the terrain of game development is flat?" Integral to this new global economy project is that I present the lives and work of India- and US-based developers as connected, for at its core, that is what the new global economy is about. These connections, often so numerous and interwoven that developers fail to recognize them, supersede talk of "our jobs" and "their jobs." The industry itself and those within it need to learn how work gets done so that laborers do not fail to grasp the importance of what has transpired. The realities of everyday developer life are connected to a much broader set of rules and ideologies.

This text lays bare the work, rules, and other unspoken realities of the videogame industry in order to ask questions about those systems. Games are made by a wide ranging group of people, who at their core are driven by asking questions about and attempting to better understand underlying systems; and yet, they rarely do it of their own industry. Most disturbingly, these systems are becoming more opaque and reinforced in ways that most game developers ought to find abhorrent, given their proclivity and desire to get at these sorts of underlying systems. For so many, this is an activity that they are passionate about, and as such, all game developers need to ask if these rules and systems are functioning as they ought to and if they might rather see a different set of rules and systems in place.[6]

By February 2005 a new project for the PSP was already underway and I had become a regular fixture about the office and was largely ignored as "the anthropologist" in the office. I partook in conversations, asked questions, and attempted to better understand that which was unfamiliar to me. I attended meetings, occasionally with introductions to those who were not familiar with "our resident anthropologist." I drank copious amounts of their coffee and partook in bagel/donut Fridays. I interviewed them and other times simply talked to them about life and work. I became friends, attended parties, and had evening beers with some of them (World 5 demonstrates that this is an important aspect of development work). I submitted talk proposals to the Game Developers Conference (GDC) with a couple of my informants. I answered their questions as best I could when they were curious about my findings. They often talked about themselves and VV as being distinctly different from other studios, though often the concerns they voiced were the same as other game developers. Some even looked to me to help find answers to their quality of life (QOL) questions, a summons that further inspired this book.

While conducting fieldwork at VV, I saw the development and release of numerous game titles. I also saw three projects started then canceled or transferred elsewhere. Game console systems came and went. Microsoft's Xbox 360 came and the Xbox left. Sony's PlayStation 3 (PS3) came, though the PlayStation 2 (PS2) still hasn't left. Nintendo's GameCube (GC) never really arrived and the Wii has taken gamers and developers by storm, albeit briefly. The Nintendo DS (DS) came and has planted its two little LCD screens firmly into the hearts of game developers.

I gathered data from videogame-related news sites, blogs, web comics, and corporate websites. I saved every press release and every SEC document I could find that might better contextualize the arena of game development in a broader political-economic context. I first searched the US Patent Office's online system and later turned to Google's Patent Search service once it became available.

In late April 2007 after returning to the United States from India, VV released their most ambitious project to date, an endeavor that I had seen progress from concept to completion. VV had swelled to more than 175 employees and various contractors. For nearly three years I sat with the developers (engineers, designers, artists, and managers) who produce the products we simply call "videogames." These developers, at least for the players of games, often disappear behind a single name: Activision, Sony, Microsoft, Nintendo, Miyamoto, John Carmack, Will Wright, *Spiderman, World of Warcraft*, Xbox, Wii, or PlayStation. As the years passed and my research continued to examine the creative collaborative work of game development, many of my Indian informants lost their jobs when game studios closed as the US economy went into recession. Even the lure of less expensive labor that might have helped in a tough economy couldn't lure American game development studios that did not quite understand how to integrate overseas teams whose limited knowledge of and experience with developing games complicated outsourcing.

I began to see in the communities of my informants a reflection of broader shifts in creative work practice. I realized that asking questions about what had changed in the ways people work missed the key questions about how game developers themselves work. What can the everyday work of game developers teach us about globalization and the new economy? How do these activities and communities differ across national and cultural boundaries? What does the new economy mean for what work looks like? My research among game developers continues as I observe old informants from a distance, seek out new field sites to work with, and create a kind of experimental system for new modes of videogame production work. My questions have changed somewhat, but an ethnographic examination of the work of videogame production and the importance of creative collaborative practice remains at the center of my inquiries.

There is a temptation, on the part of both cultural analysts and the general public to understand or equate videogame development with software development. Many educational programs even call themselves "game development" programs and focus only on the software development (often referred to as "software engineering" or "programming") aspects of game development. Young people interested in videogames are often instructed by those unfamiliar with game development to enroll in computer science programs, where often they fail to find themselves at home. Game developers come in many flavors. Artists now constitute one of the largest segments of videogame development work. Game designers, while a smaller population than engineering teams, focus on issues very different from software development. Game development has always been a "strange mix" of artists, game designers, and engineers. To equate this milieu to software does a disservice to the work of the activity as well as the very particular technological, global, and political-economic context within which this labor occurs.

Box 1.3

```
#: SET DEMO_MODE 1
Casey: There seem to be four major categories of folks working
here: engineers, artists, designers, and managers. Am I missing
somebody?
ART_Spidey_2: No, that sounds about right. So you have probably
seen that child's toy that has a board with round and square
holes and they have round and square pegs. So the engineers are
the hard shapes. They will go into the shape they're supposed
to go into. And they like going in the right shaped hole. If
it's the wrong shape, they're not going to be happy. Artists are
more like clay. They're pretty free form, you can mush them into
just about any hole. But, after they go through they're going to
retain that shape a bit. You may have to force them through, but
they will learn it and retain some of that shape from the pro-
cess. The designers are the pegboard though. They set up where
everything has a space and how it is going to work together. Now,
just because you have pegs and holes doesn't mean that every-
thing is going to go through. That is where the hammer comes in.
Managers are the hammer. Nobody likes the hammer, but you need
it anyway. Sometimes things aren't all lined up, but the hammer
comes down and pops everyone through. Sometimes it's all lined up
and easy. Other times it doesn't work so well.
#: SET DEMO_MODE 0
```

The temptation to align games with software is fed by similarities in form between the two products. For example, *Second Life*, which appears to be a videogame and the "breadth of affordances in virtual worlds owes a great deal to [its] gameness," (Malaby 2009, 14) is, in fact, not a game (though games have been developed within *Second Life*). While comparisons are drawn between *Second Life* and *World of Warcraft* (*WoW*) because each may be an immersive online experience, *WoW* is a game, designed and produced as such. One might point to the popularity of *WoW* over *Second Life* as a demonstration of what makes game development different from providing game-like spaces of play. *WoW* delivers content produced by a large number of artists and game designers to its players. Game mechanics manage the player's relationship with the underlying system of the game and creates goals and feedback loops with which they interact. This is what prompts players to return, time and again, for their repeated interactions and "grinding" in order to advance levels. *Second Life*, on the other hand, was designed to ask users to develop all of those elements for themselves. An avatar in *Second Life* has no "level" other than that defined by what the user has created for the system. No game mechanics reward *Second Life* "players" for their activities other than perhaps the currency running through the system or the pleasure or recognition of having creating something in the virtual space. Even within their virtual account in *Second Life* developers play *Tribes*, which has artwork, models, music, levels, and underlying game systems that, ultimately, are the reason workers play it after working on *Second Life* all day. This isn't to say that analysis of *Second Life* and Linden Labs should be excluded from our understanding of games, simply that there are differences between *Second Life* and *WoW*, and certainly differences between Linden Labs and a game development studio.[7]

World 1-3: Developer's Dilemma—The Mechanics of the Rant and the Genre of Zero Punctuation

The collective unease with the *Asylum* development process and its spiraling complexity coincided with (several months into the project and at the beginning of the permanent crunch) the November 2004 publication of a blog on the LiveJournal site by an anonymous poster "ea_spouse." Some of the concerns voiced by ea_spouse echoed those of the developers working on *Asylum*. For others it was simply someone over-thinking something that was common and didn't really matter, since at the end of the day it meant they were still able to make videogames. The blog, written by the "significant other" of a game developer, voiced frustrations over work practices in the Los Angeles studios of Electronic Arts (EA).

> Our adventures with Electronic Arts began less than a year ago. The small game studio that my partner worked for collapsed as a result of foul play on the part of a big publisher—another common story. Electronic Arts offered a job, the salary was right and the benefits were good, so my SO took it. I remember that they asked him in one of the interviews: "how do you feel about working long hours?" It's just a part of the game industry— few studios can avoid a crunch as deadlines loom, so we thought nothing of it. When asked for specifics about what "working long hours" meant, the interviewers coughed and glossed on to the next question; now we know why.
>
> Within weeks production had accelerated into a 'mild' crunch: eight hours six days a week. Not bad. Months remained until any real crunch would start, and the team was told that this "pre-crunch" was to prevent a big crunch toward the end; at this point any other need for a crunch seemed unlikely, as the project was dead on

schedule. I don't know how many of the developers bought EA's explanation for the extended hours; we were new and naive so we did. The producers even set a deadline; they gave a specific date for the end of the crunch, which was still months away from the title's shipping date, so it seemed safe. That date came and went. And went, and went. When the next news came it was not about a reprieve; it was another acceleration: twelve hours six days a week, 9am to 10pm.

Weeks passed. Again the producers had given a termination date on this crunch that again they failed. Throughout this period the project remained on schedule. The long hours started to take its toll on the team; people grew irritable and some started to get ill. People dropped out in droves for a couple of days at a time, but then the team seemed to reach equilibrium again and they plowed ahead. The managers stopped even talking about a day when the hours would go back to normal.

Now, it seems, is the "real" crunch, the one that the producers of this title so wisely prepared their team for by running them into the ground ahead of time. The current mandatory hours are 9am to 10pm—seven days a week—with the occasional Saturday evening off for good behavior (at 6:30pm). This averages out to an eighty-five hour work week. Complaints that these once more extended hours combined with the team's existing fatigue would result in a greater number of mistakes made and an even greater amount of wasted energy were ignored. . . .

EA's attitude toward this—which is actually a part of company policy, it now appears—has been (in an anonymous quotation that I've heard repeated by multiple managers), "If they don't like it, they can work someplace else." Put up or shut up and leave: this is the core of EA's Human Resources policy. The concept of ethics or compassion or even intelligence with regard to getting the most out of one's workforce never enters the equation: if they don't want to sacrifice their lives and their health and their talent so that a multibillion dollar corporation can continue its Godzilla-stomp through the game industry, they can work someplace else. . . .

I look at our situation and I ask "us": why do you stay? And the answer is that in all likelihood we won't; and in all likelihood if we had known that this would be the result of working for EA, we would have stayed far away in the first place. But all along the way there were deceptions, there were promises, there were assurances—there was a big fancy office building with an expensive fish tank—all of which in the end look like an elaborate scheme to keep a crop of employees on the project just long enough to get it shipped. And then if they need to, they hire in a new batch, fresh and ready to hear more promises that will not be kept; EA's turnover rate in engineering is approximately 50 percent. This is how EA works. So now we know, now we can move on, right? That seems to be what happens to everyone else. But it's not enough. Because in the end, regardless of what happens with our particular situation, this kind of "business" isn't right, and people need to know about it, which is why I write this today.

If I could get EA CEO Larry Probst on the phone, there are a few things I would ask him. "What's your salary?" would be merely a point of curiosity. The main thing I want to know is, Larry: you do realize what you're doing to your people, right? And you do realize that they ARE people, with physical limits, emotional lives, and families, right? Voices and talents and senses of humor and all that? That when you keep our husbands and wives and children in the office for ninety hours a week, sending them home exhausted and numb and frustrated with their lives, it's not just them you're hurting, but everyone around them, everyone who loves them? When you make your profit calculations and your cost analyses, you know that a great measure of that cost is being paid in raw human dignity, right? (ea_spouse 2004)

The words of ea_spouse caused a ripple in the videogame industry, one that is still being felt, though in different ways. For a brief moment, it seemed revolutionary, though seven years later, new QOL controversies emerge, demonstrating that very little has actually changed. White papers were written and special interest groups formed, and awareness was certainly raised, but death-march crunches still occur and many developers accept them blindly. The International Game Developers Association (IGDA) encouraged developers everywhere to contemplate and begin addressing QOL issues (Bates et al. 2004). A few years later, *Game Developer* magazine published an article examining the success and failure of QOL efforts in the game industry (Hyman 2007). The IGDA appeal has since metamorphosed into many different IGDA-sponsored projects, a few of which I still participate in since deeming them necessary to my research. Most of these projects (and those who created them) aren't certain how to frame their initiatives; neither top-down (management lead) nor bottom-up (instigated by the rank and file) has gained significant traction. In the pages that follow, I will argue that the QOL problems are rooted in the videogame industry's emphasis on secrecy, closed networks of access, and use of the state[8] to discipline those networks. These conditions simultaneously enable and constrain the videogame industry's practice of creative collaborative work.

The industry noticed developers' overwhelming response to ea_spouse's words. In March 2005, at the Game Developers Conference (GDC) in San Francisco, California, hundreds of game developers crowded into an IGDA sponsored session titled "Burning Down the House: Game Developer's Rant." This is a perennial event at the GDC, known to be an (in)famous gathering of developers who "cut the shit and speak truth to power" (Davis 2006). This year was different—something interesting was about to happen, for when Greg Costikyan[9] took the stage, he delivered "The Rant Heard 'round the World":

As recently as 1982, the average budget for a PC game was $200,000. Today a typical budget for an A-level title is $5 million, and with the next generation it'll be more like $20 million. As the costs ratchet up, publishers become increasingly conservative, and decreasingly willing to take a chance on anything other than the tired and true. So we get Driver 69, Grand Theft Auto: San Infinitum. And license drivel after license drivel. Today you cannot get an innovative title published unless your last name is Wright or Miyamoto.

How many of you were at the Microsoft keynote? The HD era, bigger, louder, more photo-realistic 3-D, teams of hundreds, and big bucks to be made. Not by you and me of course. Not by the developers—developers never see a dime beyond dev [development] funding—by the publishers (and Microsoft, presumably). Those budgets—those teams—ensure the death of innovation. This is not why I got into games. Was your allegiance bought at the price of a television?

Then there's the Nintendo keynote. Nintendo is the company that brought us to this precipice. Nintendo established the business model under which we are crucified today. Nintendo said, "pay us a royalty not on sales, but on manufacturing." Nintendo said, "we will decide what games we'll allow you to publish," ostensibly to prevent another crash like that of 1983, but in reality to quash any innovation but their own. Iwata-san said he has the heart of the gamer, and my question is what poor bastard's chest did he carve it from?

My friends, we are fucked! We are well and truly fucked. The bar in terms of graphics and glitz has been raised and raised and raised, until no one can any longer afford to risk anything at all. The sheer labor involved in creating a game has increased exponentially until our only choice is permanent crunch and mandatory 80-hour weeks, at least until all our jobs are outsourced to Asia. (Costikyan as quoted in Davis 2006)

Costikyan's rant lamented the rapidly changing structure of the game industry that demanded particular modes of production favoring extensive content over innovative gameplay mechanics. What is interesting about the rant sessions at GDC is that they are actually indicative of a broader genre of commentary in the videogame industry. They are one of the most widely attended sessions aside from the keynote speeches. They have recently even created situations where developers are quoted saying quite provocative things and are later reprimanded by upset studio heads or publishers. Regardless of repercussions, the rant serves as productive mode of critique in the videogame industry and among game developers. Though a rant's tone may come off as sardonic or flippant, it is a product of passion and interest. One cannot really rant without caring.[10]

More recently, the video rant has emerged as an over-the-top and increasingly vitriolic form of the rant. The online magazine *The Escapist* publishes weekly issues with game and industry commentary. In August 2007, it posted a video called "Zero Punctuation,"[11] created by the character

"Yahtzee" (Ben Croshaw). He, now famously, delivers weekly rants on what game developers have done right or primarily wrong in newly released games. Though his highly critical, often vitriolic, and seemingly misogynistic and homophobic diatribes lampoon games, they also offer needed critical commentary on the state of the videogame industry and the kinds of games being produced. His videos have struck such a chord with developers that his narratives were placed on the big screen during the GDC's Developer's Choice Awards in 2008.

Yahtzee exemplifies a common, intensely critical, public peer-review process within the game industry. Unfortunately, his words are easily criticized since they pantomime a kind of gamer narrative that uses phrases like "ass fucking" or tossing additional use of "tits" around. Yet to engage his rants at that level is as weak a reading as most games are given by "critics." Yahtzee's methods, then, place him in a position quite similar to that of the developers he takes on. Many developers comment that simply having Yahtzee publicly rip asunder their work is a sign of his respect, since his deconstructive stance requires extreme engagement and effort.

This kind of oft-heard, ranting, caustic quality to game criticism has led some to ponder whether the videogame industry is an "unhappy" place to work (Alexander 2010). I think the question is too simplistic in its approach. Many of the people who work in the videogame industry see themselves as having a significant stake in an industry in which they have little control. Game creation is a labor of love, something that developers invest themselves in technically and creatively. Developers often make it clear that they appreciate "the rant" and other extreme forms of feedback. They are not oblivious to the structures within which they all work, but despite their frustration with the industry, they continue to invest themselves in projects. This book is in part an exploration of that desire.

Ranting, and perhaps even being perceived as jaded, is also a marker of having actually worked in the videogame industry. Those veteran game developers who are the most realistic about their position in the industry and who display their love and frustration simultaneously through the rant are said to "get it." It is from this perspective that I maintain the rant can be productive. Costykian and Yahtzee do not distance themselves through their deconstruction. Rather, they intensely love and critically inhabit critically this space. As feminist and psychologist Elizabeth Wilson notes, "Deconstruction has effect by inhabiting the structures it contests. This means, of course, that deconstruction and its practitioners are always internal to and complicit with the structures they examine." (Wilson 1998, 36) Like Costikyan, Yahtzee, and Wilson, I maintain that this kind of active engagement and attention are crucial to "inhabit[ting] well" (Wilson 1998, 36).

I have always conceptualized my relationship with the game industry and game developers, game studies, and science studies as a kind of cat's cradle game.[12] The cat's cradle is "about patterns and knots"; it is a game that can be played alone, where, "one person can build up a large repertoire of string figures on a single pair of hands." However, when it becomes a shared and embodied activity through the passing of figures "back and forth on the hands of several players, who add new moves in the building of complex patterns," that the game becomes interesting. "It is not always possible to repeat interesting patterns," because of their contingency on individuals and combinations of patterns, and "figuring out what happened to result in intriguing patterns is an embodied analytical skill." Not only can the analyst examine collective patterns, but in doing so the analyst and informants will certainly learn about the kinds of patterns that they desire for the future. Players of the cat's cradle game can ask "questions about for whom and for what the semiotic-material apparatuses of scientific knowledge production get built and sustained," always searching out what was "denied and disavowed in the heart of what seems neutral and rational" (Haraway 1997, 268–269). Even in the heart of a discussion about asking who gets left out, we can leave our emphasis on another affect, yearning. I, too, can yearn for new kinds of projects, knots, and patterns.

In many respects, this book's title, *Developer's Dilemma*, comes from this critical, yet inhabiting perspective. It is, of course, a play on a classic game theory game, the Prisoner's Dilemma. In its simplest form, two players, with imperfect information are put in a situation where if one player "defects," he or she will receive the greatest personal reward. If both defect, they lose the most. And, if they cooperate, they will both receive a modest reward. Of course, the ideal solution is to always cooperate, but if you know the other person is going to cooperate, why not defect for the greater reward? Embodied in this conundrum is why, despite moments of possible tectonic shifts in the game industry, does the current structure persist? In the rant, I see the opportunity to call light to critical defections; where cooperation is thrown under the train for personal gain.

This is why the text takes on this genre of writing in boss fight sections of the text. It is done on purpose to drive home particular points in a style that is indicative of the kinds of "constructive feedback" that many developers identify with. Some may mark this language as not objective, or in extreme cases as simply unscholarly. My response is that, rather, it is both simultaneously scholarly and engaged with its subjects and those disciplines that inform it. I chose to not engage with whether the work is objective or not, as it broaches a much larger conversation regarding the foundations of social science research. This work is a perspective informed by personal experience, extensive ethnographic fieldwork, and considerable time reading field notes and news items surrounding the videogame industry through a theoretical and analytic lens that informs our understanding of what game development work is and how that can inform future social theory.

World 1-4: Learning to Use the Debugger

Each chapter or "world" of this text examines a particular component of the game development process and its scale with respect to the analytic center. Simultaneously, each chapter develops a theoretical category for understanding those processes and scales. While the text is divided in this way, connections remain between each section, conceptual categories and activities bleed together. Each world is a collection of four levels, much like a game. Each level within a world provides experiences or tools, which will serve the player in the boss fight at the end of each world. The boss fight in a game often requires a player to bring the lessons learned, or tools gained, in order to progress to the next world. Each boss fight also marks the transition from one stage of the game to the next. This process is usually introduced in the first world and levels of a game, such that the player can become familiar with the flow of gameplay. In a similar fashion each world in this book introduces you to new aspects of the socio-technical milieu of game developers.

Furthermore, each world begins with a DEMO_MODE, mimicking the way in which a game left to its own devices will often show the game in action without the necessary user input. These are done as a means of framing central issues that emerge in each level. DEMO_MODE is also used throughout the text to present ethnographic interview data relevant to the argument being advanced in that section of the text. In some cases, the arguments of the text can seem distant from the lived local practices of videogame developers. Yet, work continues all the time on game development projects ranging from the small to the massive. Continually thinking about how these issues relate back to the daily lived experiences of game developers is crucial. By framing each world with the words of developers reminds the reader that this a real space populated by workers who labor day after day at something they feel quite passionate about.

Box 1.4

#: SET DEMO_MODE 1
AUTHOR_DEV_DILEMMA: I played the game industry game, for a while. That experience took me from being a computer scientist at NASA's Jet Propulsion Labs in Pasadena Southern California to being an engineer at a small videogame company in La Jolla near San Diego. I went from developing cross-platform 3D libraries for Silicon Graphics Workstations, Linux, Windows, and Mac OS computers to developing cross-platform 3D sound systems for Windows, Mac OS, PlayStation, and Nintendo 64. Eventually I meandered my way back to engineering special effects systems on those same platforms, but only near the end of my time at this particular company.

As the dot-com bust decimated clients throughout Southern California, those of us with "other options" were encouraged to pursue them. I made an appeal to a college roommate and friend whom I'd helped land a job: return the favor. That worked out for a few more years until corporate dysfunction sent me seeking other options. There were other opportunities aside from graduate school, but the line "Where does the bastard child of a forbidden tryst between computer science/mathematics/technology and sociology/women's studies/philosophy turn?" landed me a spot in Rensselaer Polytechnic Institute's Science and Technology Studies Department.

My time in the game industry was something I seemingly forgot until chatting with DESIGN_LEAD_1 at a party early into my time in graduate school. DESIGN_LEAD_1 was in the middle of a big project and things weren't quite going as planned. There were numerous parallels between his situation and those that prompted me to pursue graduate research. In retrospect, it was probably the first of many "stitch and bitch" sessions that DESIGN_LEAD_1 and I had over the years, reflecting on growing up gamer and finding ourselves in companies attempting to create the things we love. It was also about earning the respect of our peers for the massive amount of labor and love that goes into putting boxes on the shelves of stores like Best Buy and Walmart. Those boxes that prominently feature the logos and names of the companies seemingly least associated with creating games.

Whatever prompted DESIGN_LEAD_1 to speak to ENG_GRP_MGR_1 about allowing a social scientist among the busy developers of their company is a mystery to me. I count myself fortunate, because without their willingness to try something different, my research and this monograph would have never occurred. Yet, I continually fear that it will never happen again, that game developers or publishers are simply too busy or too secretive to allow field site access for an extended period of time. But that is precisely what the game industry needs most. Thus, this labor and love, for the developers in an industry that so desperately needs it.
#: SET DEMO_MODE 0

The structure of this book is in part performative. It mimics the level structure of the original *Super Mario Bros.* game published by Nintendo for the Nintendo Entertainment System (NES). This is done to reflect the importance that recognizing discrete references to other games serves in game developer culture. Most of the developers I spoke with in this project came to games after or grew up as a part of the Nintendo Generation. This sense of a shared history and experience provides foundations for how videogame developers talk about their occupations. Of course, this is not really any different from other disciplines or environments of production where experience and language become entangled in ways that prevent broader accessibility. To make the work experiences of developers decipherable, the book is structured in a way that provides readers with the tools necessary to disentangle the unfolding narrative of game development work as it progresses.

I compare this approach to using a debugger, a software programmer's tool that allows them to observe the execution of a program's source code as it progresses. Part technology, part mindset, part tedium, and part intuition, debugging exemplifies what makes game development challenging. A debugger also allows the reader to "step-into" functions, moving from higher levels of abstraction to lower and lower levels until reaching the assembly code, which is fed to the processor. The setup gives the reader the ability to "step-out" or move up a level of abstraction after moving lower. For this text, this means starting at the level of work practice and stepping into those other functions that run in the background.

World 1 provides an introduction to the structure of the text and its over-arching arguments. World 1 also introduces "the rant" as an important conceptual category for analysis and inclusion into academic writing. Despite it being an important genre of speech and writing for game developers, the rant is a highly connected form of writing evidencing deep involvement in the topic about which the rant is written. One must be engaged and inhabit to rant. In some respects then, this commitment to intense scrutiny and attack is rooted in care: care for the industry and care for what it produces results in a kind of highly committed form of critique. This implicates the research in game developers' commitment, though this kind of reflexivity is already important to and crucial in many of the locales where STS scholars now find themselves embedded (Latour 2004). Rather than attempting to hold videogame developers and the videogame industry at a "critical distance," this work seeks to critically engage it in debate in a vernacular it expects, understands, and respects.

Broadly considered, World 2 and World 3 focus broadly on what is termed in the videogame industry "preproduction," which is the time and space where a game is first conceptualized. It is a time of contingency and system building where many elements are in flux. World 2 examines the individual disciplines that make up the majority of videogame production: game design, art, and engineering. The primary theoretical category under examination is "underlying systems and structures," and to better indicate what this means for game developers, a redefinition of "hardcore gaming" around the notion of "instrumental play" is developed. Game developers cultivate a central desire to understand how games tick, how hardware functions, and how to leverage software systems to produce interesting and innovative creative works. Instrumental play as a conceptual category links the activities of a sub-category of game players, "power gamers," (Taylor 2006a, 72–73) to the work of videogame development. In other words, there are traits that appear "gamer" in game developers, but those skills have very little to do with actual gaming. These traits are a very particular subset of gamer skills. They are the logical skills rooted in finding connections between what one perceives and the way things work at lower and lower levels of a system.

World 3 looks further forward down the temporal timeline to where experimental systems are being constructed to enable the second major stage of game development: production. Experimental systems serve as a useful theoretical framework for understanding and appreciating the ways in which game development tools structure and enable the creative collaborative practice. This section of the text does some historical positioning of tools development and the recent rise of the new (inter)disciplines of tools engineer and technical artist. World 3 also examines the notion of the "pipeline" or social and technical process by which items from artists, designers, and engineers are processed assembled into a videogame. The pipeline, one of the most important aspects of the game production process is also one of the least examined in any form. Thus world 3 devotes itself to the pipeline and its newly anointed stewards. Theoretically, World 3 examines the emergence of "creole" (Galison 1997, 46) or "faultline" (Traweek 2000) professions within the videogame industry. The tools engineer and technical artist have emerged at particular disciplinary interfaces between artists, engineers, and game designers. These individuals have been largely responsible for the development of the experimental systems (Rheinberger 1997) that are then used to create videogames. Thus, these experimental systems lie precisely at the interface between disciplines and offer both promise and peril for developers as they attempt to build game systems that cross these divides.

Production is the next major phase of the game development process. Worlds 4 and 5 empirically focus on this aspect of game development. Production is ideally the point where many of the variables have been removed from the development process and the goal is to produce all of the elements that constitute the game (the creation of numerous art and audio assets, level designs, mission designs, and the further development and optimization of game code). Production is the longest segment of the game development process, occupying a majority of the game development timeline. World 4 examines how "interactive" tools, often those developed in the pipeline, are used in concert with other established tools for artists, engineers, and designers to develop a game. The importance of interactivity and rapid feedback loops is examined in this world. Ultimately, however, the interactive process of game development also requires a great deal of social and technological assistance in the form of synchronization and build processes. These ensure all moving parts come together into integrated wholes. The elephant in the room, when one examines game production, is that all of the processes leading up to this point are far from stable and developed, often resulting, instead, in "crunch," or intense and extended periods of socially mandatory overtime, and a seemingly perpetual startup environment for game development companies. World 4 advances the idea that interactivity in this context has impacted how videogame developers view and understand their work: interactivity as a theoretical category that has significant implications for how game development work unfolds. Systems layer on top of systems and the negotiated aspect of design (Bucciarelli and Kuhn 1997) becomes a process of intense interaction between worker and technology where developers attempt to understand the relationship of small aspects of work to a large and often very complex whole.

World 5 examines the ways in which these carefully crafted systems can in some cases collapse around the teams that labor so intensely to keep them working well. This world also examines recent changes initiated in game development practices as an attempt to mitigate some of this propensity for breakdown. Of course this perpetual startup system owes some of its dysfunctional underlying systems and structures to those beyond the reach of many game developers. The text therefore moves upward in scale of analysis to examine some of the systems within which these machinations unfold. "Autoplay" (Schüll 2005), "crunch," and their relationship with desire is the primary critical theoretical category explored in World 5. The game industry has gained a reputation, at least internally, for chewing up and spitting out young excited workers. The text will show that the ways in which game development work intersects with desire, passion, chaos, and other external forces results in a perpetual startup machine.

The game industry is enormous, though communities of game developers are actually quite intimate. This seeming disconnect is a product of the massive structures of videogame publishing companies, console manufacturers, and distribution networks that ultimately make the rest of the industry function. Worlds 6 and 7 zero in on these systems through the lens of actor-network theory. World 6 is based in part on the framework developed in a previously published essay that traced the historical origins of this system of publishing, manufacturing, and distribution (O'Donnell 2011a). Rather than focusing on the historical origins of the system, World 6 examines some of these elements as they stand currently and of the social and technological systems that govern them. This section's empirical focus includes elements such as licensing, game development kits or "DevKits," and the agreements that support these practices. Through the lens of actor-network theory, I develop the notion of the inter/intra-actor-network to emphasize the importance of paying attention to how actor-networks can be structured to ensure very rigid demarcation between those on the inside and those on the outside. World 6 advances further the notion of the "actor-intra/internetwork," as a more nuanced understanding of actor-network theory (Latour and Woolgar 1986; Callon 1989; Law 1989) in the context of videogame development. The videogame industry depends, in part, on the distinction of privileged networks to delineate legitimate developers.

World 7 examines how these actor-networks are disciplined and maintained and the inter-connections between the broader videogame industry and the state (rather than assuming the actor-networks are emergent as many might claim). These inter-connections include ways in which patent, copyright, and even the ability to incarcerate are leveraged to ensure the stability and seeming "closure" of these artifacts. This world also examines how, despite efforts to the contrary, a system's users continue to leverage the systems themselves to ensure that practices exceed boundaries and to consistently introduce new points of instability. World 7 interrogates the state as a conceptual category, capable of disciplining subjects (Smith 1999). The idea that particular forms of "domination" are more "dominating" than others is important. In light of the intense boundary work being done in the game industry (and examined in World 6), the use of the state as a means of actively and in some cases violently enforcing these boundaries is critical.

World 8 synthesizes much of the text, consolidating and presenting the material in the format of a vertical slice of a game, in the form of a game-play narrative followed by a game design document. These provide the reader with a feel for the game play, description of the design, rules, and play of the videogame industry game. Both the language and artifact contribute to an improved understanding of the ways in which work practice becomes entangled with broader forces. This World performs two roles within this text. First, it clearly denotes and re-examines the numerous systems and structures discussed in the text. Second, it demonstrates an important form and text within the process of game development.

Games are a useful means to understanding complex systems, because the overall system must remain in focus, rather than just the local or the global level. There is a feedback loop between the system and the local conditions; they are inextricable. Games also provide the opportunity for players to feel out those systems for themselves, perhaps determining their own interventions or conclusions that are ultimately the skills integral to videogame development work. Put simply, World 8 puts to test the idea that by expressing the rules or design of the game industry *as* a game can have a persuasive effect (Bogost 2007). Most important though, the world's mechanics (Sicart 2008) get at the interconnected complexity of this system that is videogame development work in the context of the global videogame industry. It is a highly connected system with numerous feedback loops capable of unpredictable results and outcomes. Thus, World 8 attempts to express the enormity of the context videogame developer's work within the day to day.

This does not necessarily signify that the worlds simply build on one another, but rather that they are intimately connected. Aspects of each world constantly connect with and impact the local activities of videogame developers. It is with this in mind that World 8 returns to the empirical material of earlier worlds in the form of a game design document. The game design document is an important genre of writing for game developers that expresses well what actually underlies games. While all games to a greater or lesser degree contain narratives, all games contain extensively designed systems that structure the play space.

The text's focus is the system—creative collaborative work practice and those things that (dis/en)able it in the context of globalized videogame development work. This system runs on the ability and the desire to get at underlying social and technical systems and structures, thus this text serves as an example of the very phenomenon it indexes. It is dependent upon and produced via new modes of collaborative practice, while demonstrating the importance of being able to drill down into the subsystems that make up the game industry code.

World 1 Boss Fight: Ready? Fight!

New economy work, exemplified by game development practice, is dependent upon and producing new modes of creative collaborative work practice. The way these practices play out and the structural conditions they play out within, however, simultaneously undercut creative collaborative practice. *Developer's Dilemma* connects the diverse forces and activities—laws, technologies, and workplace cultures, for example—that make creative collaborative practice central to the way the new economy works. These same forces and activities are also capable of undermining collaborative practices. At the core of creative collaborative practice is the ability and necessity of being able to play with and get at underlying systems: technical, conceptual, and social. When access to underlying systems is undermined, so too is creative collaborative practice.

The text's focus continually returns to creative collaborative work practice. Linking up work practice to the structures within which it is situated is the best strategy for understanding why work looks the way it does. The components of work/play and interactivity that emerge from my field site work do so in relation to inter/intranetworks and a rapidly corporatizing state. I link work practice to broader structures because the day-to-day realities of work practice leaves my informants with little time or opportunity to better understand their context. The focus is continually at the local level, namely, fighting fires to keep work moving forward. While SEC and patent documents may seem distant from the everyday experience of work for developers, they are actually far closer than most realize. The connections between the local and the broader system are crucial to understanding the entire system, or game. As Dorothy Smith writes about "ruling relations," it is the "heterogenous extra-local that organizes the local" (Smith 1999, 73). It is too easy to get lost in those everyday realities without connecting them to broader structures, often leaving workers frustrated and disenfranchised.

Many organizations have also begun to encourage more playful and experimental workplaces. One indicator of this is Google's appointment to *Fortune Magazine*'s cover for the number one position among the "100 Best Companies to Work For." It is presented as a place where workers "can climb, play beach volleyball, lift weights, and go for a dip—without ever leaving work" (Lashinsky 2007). This Googlefication of the workplace has long been seen as a strategy of startups, technology companies, and videogame companies. The distinction between what is work and what is play blurs and frequently displaces many other aspects of worker's lives. Work practice becomes inevitably intertwined with another core category: work/play. This conflation plugs into a different set of drives, which enables and encourages workers to push harder and longer than they would otherwise. It also encourages them to forge new connections and think creatively. These new modes of work practice are simultaneously crucial, yet capable of being pushed to too far, dissolving into destructive work practices.

Burrowing into work practices within game development, this text will also explore how rising levels of interactivity go hand in hand with the decreasingly hierarchical or "flat" organization that has been touted as a distinguishing aspect of work in the new economy. Interactivity allows workers to experiment with the systems they both work within and create. Interactivity goes hand in hand with the connections between disciplines and cuts to the heart of what makes workers able to produce. This interactivity can also push too far, resulting in "infinite" meetings, emails, instant messages, and other forms of feedback and response. Interactivity can supplant the work, which then only gets done after hours.

But just as the industry pours its energy into interactivity, it can be blind to its interactions with the outside world. Policy makers and lawyers have not shied away from attempting to capitalize on the explosive growth of the videogame industry,[13] attempting to both simultaneously entice game development companies to their cities and to censor or penalize companies for producing game content that is deemed nebulously too violent or extreme. Lawyers have encouraged legal cases against companies for doing "harm" and to encourage expanded use of patenting and litigious action against one another and game users. This project demonstrates policy's lack of foresight and inability to adjust to new contexts, especially in highly technological industries like the videogame industry. It also points to specific locations where changes could be made to encourage mutually beneficial outcomes.

This book hacks many of the disciplines that birthed it. Numerous disciplines—communications, social psychology, psychology, and media studies—have begun to stake out videogames as their new territories. Often, a single-minded approach on game spaces and content prevail. Games are seen simply as virtual environments to study within or for new media to study. The newly emerging discipline of game studies most explicitly suffers from this myopia, not stopping to wonder about the broader networks within which its newfound demand is being produced. This project proceeds with the assumption that videogames are media and technology, both of which are constructed within extensive networks that have largely been ignored. The secondary task of the project is to place work practice in a contextualized setting, such that it comes into connection with the structures that affect and shape it so dramatically.

Preproduction: Muddling Toward a Videogame

World 2: Teasing Out Underlying Systems and Structures

Box 2.1

#: SET DEMO_MODE 1
AUTHOR_DEV_DILEMMA: Game developers often talk about the "90/10" rule of game development. The idea is that the last 10 percent of a game feels like 90 percent of the effort. The work of making games requires developers to assemble a host of ideas, art, tools and technologies in effort to "find the game" within an idea. Because of this, the process of game development is quite iterative. It is refined over time. During the early stages of game development, this process can seem very ambiguous, as developers attempt to provide as much room for that creative exploration as possible.
Casey: Are there parts of the game development process you like more than others?
ENG_Asylum: Well, there are different aspects of each part that I like. I mean, I seem to always like the part that I'm not in. I used to not like the end of projects, but now I love the end of the projects, because I'm nowhere near the end of my project.
Casey: So what do you like about the end?
ENG_Asylum: Well, it's kind of how things start coming together a lot faster at the end. You can see improvements from day to day. It's also that it's something, actually close to being a game at the point, and it is really obvious whether what you do is right or wrong at that point. You'll be fixing bugs and you'll know if it is fixed. But, when you're just first implementing something, you don't know if that is going to turn out or it often involves assets that you're not in control of. Yeah, and preproduction. I think I'm too close to preproduction right now, because I don't like preproduction. I used to like preproduction, but then I went into preproduction and it's basically a lot of documentation, and I'm not the best of writers, though I tend to write a lot anyway.
Casey: So what is it you dislike?

(Continued)

ENG_Asylum: Preproduction. No, seriously. I think the main thing is that no one makes–you don't really make many decisions. You come up with a lot of options and then leave the options open until you're forced to take a certain direction or the game dictates that it is one way or another. I mean, if I was the only person working on a game, then I could just make the decision right there, but in preproduction you can't do that, so there is just a lot of processing overhead.
Casey: So are you making an educated guess? Or are you just supporting a lot. . .
ENG_Asylum: That's the other thing. You don't really do too much implementation. It's trying to decide what systems you will use and most of the time you can't even pick those until more of the vision has been nailed down. It involves a lot of prototyping, which probably won't end up happening anyway. I've worked on several games in preproduction that never made it into production. That is kind of discouraging, when you're like, "it's pretty cool," and then you shelve it and move onto the next prototype because you couldn't sell it, or whatever.
Casey: How many times has that happened to you so far?
ENG_Asylum: At least three, probably more like four.
AUTHOR_DEV_DILEMMA: This also means that often times developers spend significant amounts of time working on games that may never actually be fully developed. Parts of a game may be designed, only to find that another studio has received the contract, the project has been shelved, canceled or were simply deemed financially risky. Given the importance of credited "shipped" titles for game developers, this can lead to developer's concern that they may work forever and never gain the industry "cred" necessary for success.
#: SET DEMO_MODE 0

World 2-1: Talking the Game Design Talk

Early phases of preproduction—that phase of game development where artists, designers, and engineers define the foundations of a fledgling game—often referred to as "concept," are so common that they can often become invisible. The leap from concept to fledgling preproduction team and then to production is a process that is rarely discussed or documented; yet the process of preproduction happens all of the time. During my time with VV, I directly observed more than four games enter preproduction and only one of those games proceeded on to production and completion; the others were canceled after varying amounts of time. Numerous other projects were also in various stages of preproduction and production at VV throughout the time I observed the preproduction and production of *Spiderman 3* (*SM3*).

While secrecy explicitly surrounds numerous aspects of the videogame industry, the process of preproduction is not a closely guarded secret—a "black art"— as much as it is foundational in a way that defies discussion among many developers. It is just what you do. You come up with an idea and you try to make enough of it to find out if it is going to be worth continued development. In the words of my informants, it is "the creation of options" (Informant and O'Donnell 2006b). Decisions are delayed, yet something must be created to determine if there is even a game worth making hidden away within the concept. The entry into preproduction is an important one because it represents the authorization to begin constructing the system that becomes a game. Yet as part of a secretive industry, not much of the preproduction process is open for outside view.

The industry's pervasive secrecy, though it does not veil preproduction itself, is in some respects an attempt by developers and the industry to hold themselves apart, as distinct from other industries. It lends game development a mystique or desirability. The idea that "This is not work like other people's work. This is not *real* work or *ordinary* work: This is *game development*" pervades the culture. The presentation of game development work as hard, but separated as though it were an intermezzo or interlude, provides it a kind of cachet. Nearly every conversation with game developers begins with a disclaimer, "not that I (we) represent the industry more broadly," or "we do things a bit differently here from the other studios you have probably visited." This insistence on difference seems cultural; it allows developers to mark themselves off from one another, as well as from other creative industries. In many respects fieldwork in game studios illuminated that self-defining as different was the demarcation of specific realm of play, not unlike a game. As cultural historians and scholars of "ludology" have noted in their examinations of play, all playgrounds are "marked off beforehand" providing the grounds for a new "absolute" order "into an imperfect world and into the confusion of life [play] brings a temporary, a limited perfection" (Huizinga 1971, 10).

While not always intentional, the secrecy that surrounds the daily work of game developers has dramatic consequences for those who choose to work in these communities. Sociologists of new economy work and of new media workers have noted that the "general lack of formality" may excite and entice young driven workers, and the seemingly "creative" and "autonomous" aspects of the work can lend themselves accidentally to secrecy (Neff et al. 2005, 321–330). Autonomous and creative work is rarely discussed within a game studio. One is simply expected to produce, either in concert or on their own, the necessary components that then plug into the system.

There is a mythology that surrounds game development (and design even more so), which is only made more pervasive by popular culture depictions in movies, or the hype that often surrounds particularly well-known designers or producers. Many imagine that game development looks like playing games and that all developers are famous members of a kind of game development "band." As my informants noted, this perception is wildly inaccurate.

Box 2.2

#: SET DEMO_MODE 1
Casey: There is a myth that being a game developer makes you a kind of rockstar . . .
ENG_DS_Spidey_1: That's a total myth. That is such bullshit. It's not a rockstar job. Rockstars are surrounded by women, first of all. Rockstars have a lot of money, second of all. I don't know how it's a rockstar lifestyle.
Casey: It's perceived as one, especially by kids.
ENG_DS_Spidey_1: I guess there is a perception. I used to have that perception. Now I wonder, "why did I have that perception?" I have no idea. I think that people have a misconception of what a developer is, or what they do, and how they picture game development. They probably see it as sitting in front of a game console, magically raising their hands [and] their creation leaps from their brain into the game and it's great.

Living in this version of Wonderland, and that's just so people know, it's false. Because you have to do work to make things happen. Sometimes it's a lot of work, and it's often delayed. I think people perceive the process of creation from the outside to be instantaneous and free and wonderful. In fact, it is work. I mean, people think about being a painter or an artist and just sitting there with their palette and going buh, buh, buh, buh, buh, and there's a beautiful picture. Wouldn't that be great? And it just isn't true. You ask people to paint and they'll say, "god, you know, I'm just really frustrated sometimes and it's hard."
Casey: Like, film perhaps?
ENG_DS_Spidey_1: Exactly. It's just not that easy. Same thing with games. Film is probably a perfect example. I mean everyone probably has a picture of what it's like to be an actor or director and seeing this wonderful movie. But, then you look at the actual process of movie making and it's very painstaking and exacting and miserable; there's a lot of work.
#: SET DEMO_MODE 0

Outsider perception of game design as outrageously fun and easy play is a myth heightened by the industry's many curtains of secrecy, which play out in different ways among individuals, disciplinary divisions, corporations, and inter/intra-corporate entities. The importance of being an insider and knowing those unknowable aspects (the "Konami Code" for example) denotes inter/intra-disciplinary or professional/hobbyist/independent distinctions. Secrets matter among engineers, engineers and artists, managers and leads, studios and publishers, even manufacturers and publishers. Secrecy crosses scales—from the level of the individual to the level of the company to the broader industry—and the secrets of console manufacturers matter for artists' and engineers' everyday activities. What worked on one project or for one console suddenly is discouraged or completely impossible in another context. Developers may not even be aware of this until significant labor has been invested in pursuing what was perceived to be the "normal" way to approach a problem. Secrecy propagates mistakes throughout the everyday working worlds of developers.

One of the most fundamental dimensions of secrecy within the creative collaborative work of game developers is the importance of "speaking the language" of the industry, videogames and videogame development. Language is a precursor to many of the other barriers to entry or "secrets" of the game industry. If you cannot access and understand the language of those who work in the game industry, you certainly cannot play the metaphorical game of game development.[1]

One bit of game developer lexicon is rooted, as is this text's structure, in a pervasive videogame industry reference to *Super Mario Bros.*, which was released in October of 1985 for the Nintendo Entertainment System (NES) (Nintendo 2003). *SMB* is a staple in the gamer vocabulary, as well as a temporal reference much like the Gregorian calendar's AD or BC. An overwhelming majority of game developers have an extensive history of playing games,[2] and many of these game developers began their gaming lives playing with the NES console. Though it wasn't the first videogame console introduced in the United States, nearly every developer talks about the impression that the NES made on them. An online email discussion with an Indian informant reveals an oft-echoed nostalgia for the NES, especially with respect to the "Konami Code" (a series of movements on the game controller that provided the player with extra lives[3]), which was first found in the game *Contra* on the NES:

Oh! Those were *the* days.
Circa 1988–92, at my hometown in India, we used to get the game console and "2 free" game cartridges for Rs. 20 per hour (approx $0.50) and then the prices came

down to Rs. 5 per hour and Rs. 25 for the whole night (evening to next morning).
And then those "5 games in 1" cartridges were available at extra cost.
Super Mario Bros., Contra, Donkey-kong, Popeye, Road Rash?
sigh (Informant and O'Donnell 2006a)

Games dominate the language of both work and play for gamers and game developers alike, but this is not a mechanism to keep others at bay or explicitly exclude. This vernacular works because games provide discursive resources for developers trying to describe abstract concepts, like game mechanics. Because there is no "discipline" of game design or game development, games themselves have become a kind of *lingua franca*. And when you think and talk through/with games, they become aspects of the workplace.

One particular experience while in the field demonstrated the importance of this vernacular for developers. The term "Vertical Slice" (VS) loomed large in the discourse of game development when I found myself among the developers at VV. Like most outsiders, I had never heard the term before, in part because I had never worked on a project that had reached the VS stage. VS for game developers is having one example of everything that would then go into the game: examples of every special effect, every game mechanic, and one feature-complete sample level. It also signifies a kind of barrier between preproduction and production. Because a VS has the potential to demonstrate each feature or mechanic of the game, it means that many of those "options" mentioned in the opening DEMO_MODE have solidified.

When deadlines loom (and seemingly more so around VS) more meetings crop up. It seems inevitable that just when developers want to spend more time at the computer working on the game, more meetings are scheduled that in turn keep them away from their "real" work. For this reason, many of my informants hated VS, it was a necessary requirement, but the production of it meant numerous hours spent in meetings and time spent late at work attempting to then implement the decisions made in those conversations. In one particular VV meeting the project leads and upper management were attempting to determine the mechanics of a potential Wi-Fi multiplayer aspect of the game.[4] Of course, meeting participants were already exhausted from recent long nights, and the thought of having to define a new game mechanic and underlying technology to support it was low on everyone's priority list. Engineering was saying one thing, design was saying something else, and management yet another. From my seat, it seemed like they were all saying the same thing, but the meeting continued for nearly an hour before one designer looked at an engineer and asked, "So, do you mean it's like *Spy vs. Spy*?"[5]

After this suggestion, one of the engineers thought for a moment, and said, "Yes, like *Spy vs. Spy*." The managers and the rest of the designers in the room also nodded their heads. So, similar to the language uses noted by other anthropologists who examine knowledge work, game talk and game development talk accomplish numerous tasks for game developers. Insider language, as Traweek (1988) notes, "creates, defines, and maintains the boundaries of this . . . community; it is a device for establishing, expressing, and manipulating relationships in networks; . . . it articulates and affirms the shared moral code about the proper way to conduct [scientific] inquiry" (122).

Game talk, at its core, appeals to an almost instrumentalist logic. "Like *Spy vs. Spy*" is actually getting at a deeper understanding about the mechanics of a game. The talk appeals to the game and its underlying systems in a fashion that gets at not precisely the content of the game, but its functionality. But while game talk can be a productive tool for uniting disparate disciplines (a topic covered in more depth in World 3), it can also be used to exclude. Anthropological studies of high-energy physics have shown how oral communication enables collaborative communities, but also simultaneously can be used to close communities off. These findings hold true of game development as well.

Figure 2.1
Spy vs. Spy for the NES

> Access to this world of oral communication is quite limited. In a community with easy access to widely disseminated written information, keeping crucial information accessible only in oral form is an impressively effective means of maintaining its boundaries. . . . Protection of oral communication encourages the development of a closed community. In physics it is consistent with the group's image of itself as a meritocracy: only an informed, worthy member of the community will know what is to be said and what is to be written. (Traweek 1988, 120)

Oral communication, though, is not used to maintain boundaries simply through resisting documentation. It is also used as a means to convey information for which my informants had no other language. It is, therefore, reductive to consider oral communication only a means of exclusion, for the use of game-talk serves productive capacities crucial to the collaborative capacities of developers. It is incredibly consequential that access to oral communication is closely controlled. Anthropological studies of technical work practice have demonstrated that the closed access to oral communication inhibits our understanding of what goes on in the workplace (Traweek 1988). This has consequences at the level of the corporation, but also in the frequent assumption that game development is "merely playing games." Secrecy simultaneously is used to control access to the community, but it is also productive. It provides the foundations for productive conversations across disciplines. Insider language also does work for those that use it. It is not "simply" jargon. Embedded within insider language is a greater depth of understanding and knowledge than is abstracted into what is often construed as game-geek-technobabble. Given that argument, we can see why Orr (1991) might find that "one of the interesting results of the ethnographic investigation of work practice is that one discovers that what is done on the job is often rather more than and different from the job as described by the corporation" (12). This makes sense, given that oftentimes the work is disguised by language that may be inaccessible to the "corporation" but serves important roles for workers.

Because game development work practice is quite different from what it is thought to be, and because developers do not communicate actual practices outside of their small communities, it becomes difficult to learn from and understand the experiences of others. While companies continue to invest in ways of attempting to capture this informal and abstract information that surrounds daily practice, these efforts remain problematic precisely because of their roots in social relations in addition to technological ones (Hakken 2000b). Because the social and the technological are so tightly intertwined in these interactions, it becomes tempting to "record everything," though this rarely results in a greater ability to make sense of the subsequent gluttonous data flows that are often ignored when the next project hits. Yet, it is precisely this information that might, if shared more broadly, enable companies to make sense of the data as analysts and hobbyists make sense of it for them.

Most aspiring US developers have become accustomed to needing to learn the language of game development on their own without tutorial, manual, or mentor. For example, being able to understand that when someone says, "Like in *God of War*'s context sensitive button-pressing minigames," that person is referencing a game for the PlayStation 2 console and a particular game mechanic that requires the player to time precisely certain button presses for unique or context sensitive actions. This longer statement is generally shortened to "quick time events" or "button time events." It is this complexity and contextual character that makes the emergent game design discipline particularly difficult.

The discipline of "game design" is relatively new among game developers, but it has quickly become the professional aspiration for many young game developers. This is likely because designers are on the front line for constructing what is finally viewed as the game. While people new to the industry imagine that engineering is the work of game development, it is frequently the game designers who occupy the privileged position as "author." The famous designers of the game development world—Will Wright, Shigeru Miyamoto, John Romero, Peter Molyneux or Cliff Bleszinski ("Cliffy B") typically lead an army of game developers, artists, engineers, and other designers who construct the products (which are then credited back to their generals). While design has long been a task among those making games, the specialization has been relatively recent. Many designers were previously engineers or artists who have transitioned more exclusively into the designer role. Each designer I met came from a different background—for example, physics, computer science, media studies, film studies, graphic arts, writing, and journalism. More designers in each studio were "self taught" than were artists or engineers. While designers seemed to come from every disciplinary background imaginable, the common theme was that they were more likely to be gamers than members of any other discipline within game development. Designers frequently had skills that seemed to transcend disciplinary boundaries, including analytic skills that allow them to deconstruct games, examine their core elements and mechanics, and determine the underlying rules and structure of a game.

Engineers, in particular, recognize the difficulty that the design process of a game presents and appreciate the analytical approach of game designers. In part this is due to engineers having to work closely with designers to build the systems that designers then leverage throughout the game development process. Engineers recognize good designers, but too often there is no attempt made to leverage that knowledge.

Box 2.3

```
#: SET DEMO_MODE 1
Casey: So what is it that makes game design particularly
difficult?
ENG_DS_Spidey_1: The task of the game designer is incredibly
ill-defined and there's no way to teach a person how to do it
and there is not even a really good idea of what makes a good
game designer. But, on the other hand, I've worked with good game
designers, they do exist. Being a good designer has happened, so
we need to teach it. We need to learn how to package it, and that
isn't a question that has been addressed. No one has ever got-
ten fifty great game designers into a room and said, "What do you
guys fucking do?"
#: SET DEMO_MODE 0
```

And despite the fact that nobody puts fifty great designers in a room to ask what they do, we do know one thing: they play games. The need to play games to make games is as much about culture as it is about understanding what it is that makes a particular game interesting or fun. One must know the game, but also know how it functions. Because designers must play a significant number of games to be able to break down games into the component parts, they frequently come to speak in the language of games, rather than in any single disciplinary language, "Like *Spy vs. Spy*," becomes a bridge between an imaginary concept and an actual game mechanic. Designers are fluent in the language of games broadly defined, including tabletop games, role-playing games, board games, and videogames.[6]

In fact, the mechanics and the systems underlying games is what drove most game designers' interest to jobs in game development. Some came up through the organization as quality assurance testers; most others transitioned from engineering or art to the design teams. Some designers have taken existing videogame engines and customized them, building MODs or levels to demonstrate their abilities. Those coming out of software-heavy backgrounds may have created small stand-alone games. Those who were hired directly as designers were coming from other game studios where they had followed similar tracks through the organization. Though "game design" education programs have existed for some time, my informants, at least thus far, said there was no indication that these students were any better at design than someone who had come out of a physics program and was intensely interested in games.

Because the "science" of game design (and perhaps of design more generally[7]) is in its infancy and most game developers have had a difficult time deploying many of the ideas developed in the academy (or simply do not have enough time to implement them prior to the next deadline), most designers in my interviews, field site observations, and personal experiences in the industry expressed the urgent need for better tools or new ways to talk about and perform game design. Nebulous concepts like "play," "fun," "verisimilitude," and others abound, propelling and constricting new designs in different ways. In the formal-method vacuum, designers speak in terms of games—all sorts of games. This is different from explicit secret keeping or the lack of documentation of design practice. As game design has attempted to define itself, it has struggled with an adequate vocabulary or sets of practices by which to systematize its internal logics and methods. This isn't to say that there haven't been ambitious and productive attempts from both industry and academia to do so (Perry and DeMaria 2009; Salen and Zimmerman 2005; Schell 2008).

One of the ways in which we can talk about game design is via the concept that game development companies, like many new media companies are continually in what Neff and Stark call "permanent beta":

> The influence of design—where the design of products, technology, or services—and organizational form on each other emerges partly due to the process of continual technological change, in which the cycle of testing, feedback, and innovation facilitates ongoing negotiations around what is made and to organize making it. We call the organizational state of flux that emerges from this negotiation "permanently beta." (Neff and Stark 2004, 175)

I also see this state of flux as a product not only of the process and organization, but also of the relative youth and lack of institutional memory that these industries seem to exhibit. There is nothing inevitable about the permanent beta state. Permanent beta, is in part a product of technological change and flux, but it is also about the problematic character of then retaining those experiences or learning long-term lessons about how precisely one works in those situations.

The critical gap for would-be designers is making the leap from talking about games to constructing games. There is a cavernous gap between thinking about game design and actually doing it. Both the practicing designer and would-be designer must be able to translate that gamer vocabulary into the intermediary languages of engineers and artists. Designers must be able to write and choreograph the experience of playing a game, balancing their personal desires with what other players will find fun, interesting, intriguing, or meaningful. This means that designers frequently cross between the worlds of artists and engineers, and experimentally construct ways through which they bridge code and art through the nebulous mechanism "data," which is examined more closely in World 4-3.

The language of game design, rooted in a deep and nuanced understanding of games, provides discursive and conceptual resources for developers to draw on in their discussions surrounding abstract and often complicated systems. While game designers may locate their ideas and rule systems in the language of other games, mathematics, spreadsheets, or others, game design is at its core about the construction of systems (underlying systems) that others will play in a game. Game designers locate the core components that others will play and feel out in the process of gaming. Game designers are, therefore, gamers par excellence, for they are always looking for loopholes or means to game the system. Particularly in the early stages of game development, design takes a leading role, identifying the rules and systems, or simply mechanics, that will govern game space. Designers will frequently attempt to do things that they have been instructed will not work, to see if there are ways in which to tweak their work so it falls within guidelines. The essence of game designers is to embody "teasing out underlying systems and structures" and "instrumental play," in ways that construct worlds that players want to experience time and again.

World 2-2: Bumping into Software

Box 2.4

#: SET DEMO_MODE 1
AUTHOR_DEV_DILEMMA: Some artists pursue game development specifically. Others happen upon it more accidentally. Making artwork for games can be a very different process than other forms of graphical artistry. For some, that is precisely what interests them in the productive process.
Casey: Did you intend to go into games when you went to art school?

```
ART_DS_Ogre_1: Sort of. That is a funny story. I was really into
the first The Sims game. I discovered that there is a whole
online community devoted to editing the skins, which I thought
was really cool. I tried to teach myself Photoshop and just
enjoyed spending all day making alternate textures and wanted to
get into editing the object files and changing the animations for
the characters. I couldn't figure that out through the tutorials,
so I started thinking, "Oh, maybe there is a school for this sort
of thing, because I could deal with getting paid to do this all
day." So, yes, games was what I was thinking when I pursued art
school. I mean, I didn't know anything about games and certainly
had no idea about how they were made or what went into it.
AUTHOR_DEV_DILEMMA: Art is thus closely tied to the underlying
systems of a game. There are very precise requirements that must
be met for those "assets" to make their way into a game engine.
These are requirements that an artist must balance with a host of
other requirements in their quest to bring visual life to a game.
#: SET DEMO_MODE 0
```

It is often impossible for artists, engineers, or designers to know precisely what they can create without causing the underlying hardware to buckle under the pressure they put on it. For artists, in particular, preproduction can be a difficult time: they must frequently be the Swiss Army knives of game development, balancing between fine arts skills and technical knowledge. Preproduction is often the time when getting artwork into the game is at its most difficult because engineers have not yet determined how artwork will make its way into the game.

But artists' work is significant to the game and its requirements, for the majority of a videogame's CPU usage actually goes to the rendering or drawing of images to the screen and represents the overwhelming majority of what game players see on screen. Examples include visual interface elements like "health meters" or user interfaces (typically called "HUDs" or heads-up displays), as well as actual game content like 3D models and 2D textures. Animators put these models into motion and texture artists give them "skins." Concept artists sketch foundational pieces that define the look of a game. Full-motion video artists assemble game cut-scenes. Lighting artists, with an understanding of the dramatic effects of illumination, place light sources in maps to set a game's mood or tone. Most games dramatically redefine the user interface and rarely make use of standard user interface elements. Each of these artistic elements must be created by an artist and made available to the game "engine" in some format. The proper format for those elements or the steps necessary to then see them present in the game is often in a state of flux during preproduction. Artists may be involved in defining the flow or pipeline that artists will later use to examine their creations in the game. When what they have spent so much time creating does not appear in game or does not look as expected, artists must frequently engage with the knowledge of engineers and designers.

Generally speaking, artists are trained either in fine arts programs or more professionally minded institutions, which claim to balance artistic training with up-to-date software package training. Self-taught artists will occasionally make their way into game studios, though in my research this was the exception rather than the rule.

Much like their engineering counterparts, artists develop particular areas of expertise over time. Each must be familiar with sophisticated technological tools that are used to create their artistic visions, and these tools are highly specialized for each discipline. Even if an artist is familiar with the software package 3D Studio Max ("Max") or Maya, both products created by AutoDesk, they may not be familiar with the particular add-ons or additional applications made to run within these programs in an effort to speed or simplify particular tasks. Artists must be able to quickly grasp and work with new tools, in many cases custom technologies that may not have been designed by artists.

Artists have specialized languages for understanding their work, just as the other specialized roles do; and artists' communications are further contextualized by the tools and systems that they use to produce art assets. Often, terms can conflict across disciplines. At one level, artists and engineers speak different languages because the same words have different meanings.

Programmer: "I need this model in under 300k."
Artist: "Okay."
Artist: [Spends a week and makes the model in under 300,000 polygons.]
Programmer: [Head explodes.]

The joke is that the engineer meant one thing and the artist interpreted the words to mean something entirely different. The engineer was thinking about the size of a file on disk; the modeler interpreted the request to constrain the number of polygons in the model. This example ignores a significant difference between art production for games and art production for other purposes. The limitations of games' hardware and software systems alter the artistic process. Yet at the same time, many artists find the limitations compelling. Constraints provide a framework within which creativity can run rampant. Rather than being left with a world of possibilities, artists can explore innovative ways to work within the more limited numbers of creative possibilities available to them (at least during preproduction).

Game development really occurs at the interface between artists, designers, and engineers because conversations between disparate worldviews result in a complex creative works of artistic information processing known as a videogame. At every point of preproduction, communication from artists offer visual clarity, fidelity and creativity to complement the designers' underlying mechanics and messages in a way that engineering can assemble into a final product. While there are myriad ways in which an artist might construct, animate or texture a model, the process is always mediated. Engineering may eliminate many of the possibilities. Design may restrict how elements may look or function so everything will work with other elements of the game.

ART_DS_Ogre_1: Certain things are just harder for me, I guess. I'm very comfortable and happy doing individual tasks. Give me a list of things to make, what style you want them in, or whatever, and I will make them. But planning something out in a huge scope is different. Tile sets for example, you make these individual pieces that work for various situations. It has to work here like this and there like that. It always has to work. That's just too thinky for me. I don't like to think that much. I guess I'm just not that creative. So, I'd never done levels before and I get this sketched up file from design and it is just the platforms for the game, just boxes and it is very angular, and I'm thinking, "But this is a forest, why is that floating?" So then I have to imagine something that justifies the character being able to go up on that platform. I hate that. I want someone else to do concept and I can just make it. (Informant and O'Donnell 2007a)

As many facets of the team constrain the artists' work, preproduction demands documentation of how things are to be done. Much of the preproduction work of artists, including meetings that attempt to marry the differing desires of designers, artists and engineers, can feel very much like non-work. Though it may not be the aspect of game development that artists most cherish, all of the non-work is crucial to the later success of the project, and the expertise of experienced artists is crucial to the planning of a games production.

When in preproduction, artists who are not in meetings sit in front of their computers,[8] gather in front of white boards with other developers, talk in person with developers, and sit reviewing one another's work. These peer review sessions can frequently result in tense situations where artistic style comes into conflict with the overall aesthetics of a game. Colleagues ask, "Would he really move like that?" and press each other by questioning the quality of one another's work or challenging one another's assumptions about how a given task ought to be done. Collaboration, changes, and creative innovation is often spurred by questions like, "Have you seen the animation that the other team made?"

Artists also come into conflict about technological limitations, and they don't always react well to implied criticism. "Of course it would look better like that, but I've only got seven bones[9] to work with." They remind their colleagues about limitations under which they toil. "If I do that, I'm going to break our budgets." Entire conversations can be dominated by the examination of several discrete frames of animation or the tweaking of several bones or vertices on a model. In some cases, those animations or models may never make their way into the finished game product.

Some of the parameters within which artists struggle, beyond the constraints put upon them by design and engineering needs, are the layers of idiosyncratic software and hardware systems. During one conversation, a technical artist talked about how willing artists were to assume that if something wasn't working, it was their own fault. They would continue working with a model or other art asset, forcing it into the game, despite repeated failures. In some cases they would even manage to make something fit into the game that should not have fit in the first place. It seems quite important to artists that they fit their work into the parameters established by others, for they will continually modify and carefully check their work to see where changes can be made to improve the visual quality of their work without breaking the guidelines they have been given.

Many artists feel like work horses, doing the hard logistical work of preproduction, and while very few artists are involved in the preparatory work of preproduction, the number of artists rapidly ramps upward as a game reaches production. Those artists who were part of preproduction often look forward to entering production, as it is a time in which they can, "simply go in their hole and make some stuff." Artists contribute concretely to the production of art (or "assets") for the game, and they are rarely directly involved in the production of code or data. In fact, this distinction—art "assets," rather than code or design—have led many game development companies to seek art production work overseas because the linguistic distinction was logically differentiable to management. This partial outsourcing, wherein art is separated and isolated, neglects the connected character of artistic game work.

World 2-3: Bumping into Hardware

Game designers and engineers spend a great deal of time constructing the videogame's foundation and underlying systems. Perhaps the collaboration leads would-be developers to pursue degrees in computer science more often than art, game design, or film studies due to the assumption that games are "just" software. The cultural imaginary around code is so closely linked to computer science that they are almost synonymous, while other fields are only now embracing the role of code. Engineers (and they do refer to themselves as engineers rather than programmers or computer scientists) translate design mechanics into computer programming languages like C or C++, which are then compiled into machine understandable code by software often made specifically for a given game system. These compilers can be, to greater or lesser degrees, integrated into other game development tools.

Box 2.5

#: SET DEMO_MODE 1
AUTHOR_DEV_DILEMMA: When asked about why games command their attention, engineers often talked about how producing a game was simply more interesting than the possible alternatives.
ENG_DS_Spidey_1: Because they're games and that's where the excitement comes in. It's more fun to make things blow up on screen or make something animate, than it is to make a web polling application. . . .
AUTHOR_DEV_DILEMMA: Yet, engineers are quick to note that what is on the screen is rarely their direct doing. Their engagement is at a different level. The focus is closer to the hardware and to all of the elements that must be coaxed into a cohesive whole to make those elements animate on the screen. The process of determining what is necessary to bring the hardware and software systems into that kind of harmony, however, can be quite complex:
ENG_DS_Spidey_1: There is this total amount of knowledge when you're going all the way down to the wires and you have to understand everything from what you're typing in to what's coming out all the way down to the basics of computer hardware. Really, you know, how computers work at a hardware level, and you can infer things. But you have to have that desire to really know what's going on and make things work fast and work with obscure hardware that changes every year and a half. A lot of people don't have that, I guess. A lot do.
AUTHOR_DEV_DILEMMA: For a community that manages to burn through young talent, it troubles me, as an analyst and developer to hear how vitally important experience seems to be. Consistently informants will cite the dramatic difference in understanding that

(Continued)

```
developers acquire as they spend greater amounts of time engaged
with the process of making games:
ENG_DS_Spidey_1: There is a tremendous gap. It's just cavernous.
Between the knowledge that they teach you in a curriculum and
actually creating software and games even more so.
AUTHOR_DEV_DILEMMA: There are numerous reasons why those gaps
exist, only partially the fault of educators. Given the kind of
secrecy and contextual character of game development work, it
seems almost a given that the distance between those worlds would
be gaping.
#: SET DEMO_MODE 0
```

At the same time that they work to turn mechanics into code, engineers invest significant effort into externalizing the interests of artists and designers. They construct the means by which designers can adjust the parameters of design mechanics, even though these parameters may change dramatically over the course of game development, as the team learns more about what makes the game particularly enjoyable or meaningful. Engineers often refer to this abstractly as "data driven" design, but it's not just about honoring the data. The way engineers work also empowers designers, providing alternate methods for constructing the game system that does not always require the intervention of engineers (who might already be stretched thin by all the demands on their skills).

Websites like *GameDev.Net* that provide information for developers offer resources aimed at helping would-be engineering developers learn the tools of the trade. Some examples are *Nehe Productions, Code on the Cob,* and *AngelCode,* all of which offer extensive information on technical issues and the concerns of engineers who are learning game development. These sites' engineering-heavy focus and marked lack of information about how design, art, engineering, and management work together to produce games is not widely or explicitly discussed.

Pointing would-be game developers toward conceptualizing the practice of videogame development collaboratively would be more productive; encouraging engineers to think about the place of artistic and design practice will better serve the student and the industry. What follows, is the sample output and source code often used to introduce engineering-inclined game developers to 3D game programming. While it is a helpful demonstration, it actually confuses the roles of artist and programmer. For example, it would make more sense for demonstrations to read the "pyramid" data from a file that can be later modified to produce other outputs,

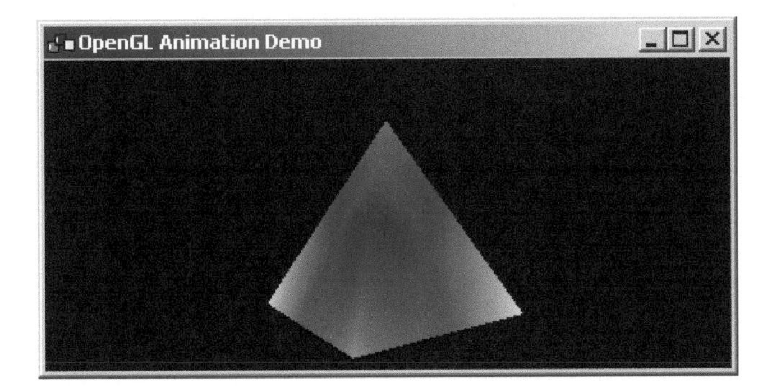

Figure 2.2
Screen shot of a simple OpenGL animation window

```
// Clear the GL Context
glClearColor(0, 0, 0, 1.0);
glClear(GL_COLOR_BUFFER_BIT | GL_DEPTH_BUFFER_BIT);
glLoadIdentity();

// Get Elapsed time in Seconds Since Last Update
double rElapsedTime = wxGetElapsedTime(TRUE)/1000.0;

// Draw Some Stuff Based upon Time... (Borrowed from Nehe's Tutorial #5)
// http://nehe.gamedev.net/data/lessons/lesson.asp?lesson=05
glTranslatef(0.0f,0.0f,-4.0f);

// If you just want it to spin, use this line.
// m_rRotationAngle += 1.0;
// If you want to update it based upon time...
// m_rRotationRate is the number of revolutions per second of the triangle.
// the default rate is being set in the constructor.
m_rRotationAngle += (m_rRotationRate*360.0) * rElapsedTime;

glRotatef(m_rRotationAngle,0.0f,1.0f,0.0f);

glBegin(GL_TRIANGLES);              // Start Drawing The Pyramid
   glColor3f(1.0f,0.0f,0.0f);       // Red
   glVertex3f( 0.0f, 1.0f, 0.0f);     // Top Of Triangle (Front)
   glColor3f(0.0f,1.0f,0.0f);       // Green
   glVertex3f(-1.0f,-1.0f, 1.0f);     // Left Of Triangle (Front)
   glColor3f(0.0f,0.0f,1.0f);       // Blue
   glVertex3f( 1.0f,-1.0f, 1.0f);     // Right Of Triangle (Front)

   glColor3f(1.0f,0.0f,0.0f);       // Red
   glVertex3f( 0.0f, 1.0f, 0.0f);     // Top Of Triangle (Right)
   glColor3f(0.0f,0.0f,1.0f);       // Blue
   glVertex3f( 1.0f,-1.0f, 1.0f);     // Left Of Triangle (Right)
   glColor3f(0.0f,1.0f,0.0f);       // Green
   glVertex3f( 1.0f,-1.0f, -1.0f);    // Right Of Triangle (Right)

   glColor3f(1.0f,0.0f,0.0f);       // Red
   glVertex3f( 0.0f, 1.0f, 0.0f);     // Top Of Triangle (Back)
   glColor3f(0.0f,1.0f,0.0f);       // Green
   glVertex3f( 1.0f,-1.0f, -1.0f);    // Left Of Triangle (Back)
   glColor3f(0.0f,0.0f,1.0f);       // Blue
   glVertex3f(-1.0f,-1.0f, -1.0f);    // Right Of Triangle (Back)

   glColor3f(1.0f,0.0f,0.0f);       // Red
   glVertex3f( 0.0f, 1.0f, 0.0f);     // Top Of Triangle (Left)
   glColor3f(0.0f,0.0f,1.0f);       // Blue
   glVertex3f(-1.0f,-1.0f,-1.0f);     // Left Of Triangle (Left)
   glColor3f(0.0f,1.0f,0.0f);       // Green
   glVertex3f(-1.0f,-1.0f, 1.0f);     // Right Of Triangle (Left)
glEnd();                          // Done Drawing The Pyramid

// Flush the GL Pipeline
glFlush();
```

Figure 2.3
The C++/OpenGL code necessary to generate figure 2.2

rather than a basic rendering of a hard-coded object to the screen. This would give engineers a better sense of how a game might actually get developed based on those samples. In creating a videogame you would not want to "hard code" every vertex (a point in 3D space) or color for every model in your game, especially considering that most models for games will have thousands of vertices. The animation and the code required to generate a pyramid using OpenGL is illustrated in figures 2.2 and 2.3 respectively. More useful for aspiring developers would be a demonstration of reading the pyramid from a file, or reading an arbitrary number of points and colors from a file.

The same argument for encouraging the engagement of engineering and design practice would be true for sites focused on artistic practice. The collaboration of artists, designers, and engineers is made evident in the way in which data is placed into the game's code. Though it seems counterintuitive, data is read from files rather than hard-coded into the code of a game. Because a designer would potentially want to edit the "m_rRotationRate" (the rate at which the 3D pyramid will rotate in space) in the above sample, it was probably parsed from design data. The model itself (a pyramid in this case) would have been created by and put into a file by an artist. It is therefore simpler for an engineer to allow the game to read data produced by artists and designers than to place the information themselves into the code for a game. This choice, which might not make sense from a computer science standpoint, is about enabling collaborative practice. Educational websites should give prospective developers a sense of this cooperative environment and its technical requirements and constraints. My informants often cited a lack of this kind of knowledge by engineers new to the industry as a critical failure.

ENG_DS_Spidey_1: I mean, I think on the whole, what we do in games programming is harder. I mean, in terms of total amount of knowledge you need to have the comfort with computers and how they work. I think there are a lot of programmers that don't really understand what's going on behind the scenes. They know what goes in and they know what comes out, and they kind of have a mapping of it in their head, but they don't really know what is going on. That is the difference with game programmers. Good game programmers, at least. A lot of people really understand what is going on and can spot things immediately in terms of underlying things. I mean, just this morning, we had a random problem and ENG_DS_Spidey_2 and I just immediately knew it was X, but we wouldn't have known it was X if we didn't have this obscure piece of knowledge about the hardware on the DS that lets us know, "oh, that's it." And it's perfectly possible to write code for a DS without knowing that piece of information. So it's the difference between people who pursue those little details and those who don't. (Informant and O'Donnell 2005a)

The lack of available information about the way in which game development engineers have to work became most apparent to me while working with a team in India who was developing a game prototype for the Nintendo DS. I had already shared with the development team most of my game development web bookmarks when I noticed one of the artists seated by the desk of an engineer on the project. The two were pouring over what appeared to be a hard-coded array of data. I asked what they were working on and they explained that they were trying to get the artist's data into the game. I asked why they were doing it that way rather than reading the files into the game while it was loading—so that the artist could make changes to his artwork and it would easily be reflected in the game. Both responded that the sample code on the sites I had pointed them toward, as well as in Nintendo's documentation, contained all of the data hard-coded, and they assumed that was how it had to be done. This type of misinformation in developer resources is made even more problematic when developers work on proprietary hardware, for code references can be even more difficult to find.

Box 2.6

#: SET DEMO_MODE 1
AUTHOR_DEV_DILEMMA: Many engineers I encountered maintained that there was a necessary special breed of programmer that wants to work in the game industry.
Casey: So, what does it mean to be a good programmer? What kind of person makes a good engineer?
ENG_Asylum: Well, first of all you have to be logical, and have some sense of organization. That is first and foremost. The other element is, well, it depends on your skills in what I consider to be the two main categories of programming. I think these two things compose at least 90 percent of what you do in programming. The two things are number-crunching and string parsing. And then there are a small number, like 10 percent where maybe you do some work like writing a data structure, but that is mostly number crunching anyway. So you either have to be good at one, or the other, or both. But that is just my bizarre theory on programming.
Casey: So how do you reconcile the making of good engineering decisions and bad engineering decisions and how those cascade down?

(Continued)

```
ENG_Asylum: So, bad decisions can be made because you don't fully
understand the problem or you've perhaps encountered the problem
before and go down one path and not another based on that experi-
ence rather than on looking at the big picture. And you'll always
have to revisit things, which is fine, but depending on how
you've decided to do something, you may have to redo it instead
of merely fixing it. That would be because it doesn't meet the
needs of the customer. As a programmer, the designer is the cus-
tomer of the programmer. So, designers will want a feature and
the programmers will implement it to meet their requirements.
Casey: But if a designer doesn't know what those requirements
need to be?
ENG_Asylum: Then that creates a bottleneck. Either you make that
feature so vast it handles everything that a designer could pos-
sibly want. Or at least a lot of things they might want. And
sometimes that is a good option, because it will be more general
for down the road. But, that can take longer and sometimes ends
up being less efficient.
Casey: So for the vertical slice last week?
ENG_Asylum: Yeah, so that was mainly supporting the aesthetics
that they wanted to add. So much of that was helping them just
get it into the engine, and how that pipeline worked. The code
was pretty much immutable up to that point, just helping them fix
up and support what they were trying to add. At that point adding
features would have made no sense, we had features two months ago
that weren't even in use yet.
AUTHOR_DEV_DILEMMA: Engineers in the game industry attempt to
manage the space between allowing for creative possibility and
managing the kind of complexity that that goes along with provid-
ing a system that exposes such flexibility.
#: SET DEMO_MODE 0
```

Despite engineers' best efforts to "scope out" a project at the beginning, based on the specifications of a system, they frequently discover there is much to understand about the underlying game systems. One engineer in particular discussed having a feel for "every transistor and chip inside that thing" after performing optimizations for the Nintendo DS handheld console (Informant and O'Donnell 2006b). Despite all of the documentation, specifications, and scoping, it required his investigation of and prying into the system to get it to do what they had assumed it would do in the first place.

An engineering group manager who had gone through this process numerous times talked about the process of digging into a system, at the level of software or hardware systems, and its requisite attentiveness. When I asked about mechanics that weren't working properly, ENG_GRP_MGR_1 noted:

> There is an emotional component of it. . . . You gotta get over that pretty quick. And I guess that's where the logical part comes in. You say, well, you know, it's broken. . . . you're not expecting a system call to do what it should, or hardware to do what it should, and I think . . . if you take the time, the gruesome horrible time, you can always, if you have to, map out the transistors, and follow the flow of logic. It will all come out in the wash. So you have to be persistent. (Informant and O'Donnell 2005b)

Some STS scholars have written about the process of debugging or feeling out systems, and Turkle (1997), in particular, discusses the process as "combining the magic of emergence with the possibility of getting your hands dirty," it was a combination of "hard and soft, bricolage and algorithm" (143). Yet, bricolage seems to go too far; it is not so un-systematic. Engineers learn just as much from "negative knowledge," knowing the limits of what they know as they do from knowing specific systems. As noted by sociologist Knorr-Cetina (1999), negative knowledge can steer us away from wasting time with, "things that interfere with our knowing, of what we are not interested in and do not really want to know" (64). Examining and determining how systems work is a simultaneously thorough/methodical and creative/intuitive thought process. To get at underlying systems, as anthropologist of medicine Emily Martin (1997) has noted, our tasks and relations with them begin to approximate "disorders" where our "exaggerated sense of urgency" and "exaggerated sense of boredom" contribute to our abilities to "stretch, cram, speed, warp, and loop poor old linear time and space" (253). The capacity to "organize the chaotic mix of seemingly unrelated simplistic elements into a more integrated and comprehensive framework of understanding, approaching a clearer picture of complexity," begins to approximate the clinical definition of Attention Deficit Disorder (Martin 1997, 254).

Engineers are not disordered, of course. They're systematic in their rooting for problems that keep their code, designers' visions, and artists' assets from working properly. ENG_DS_Spidey_1 noted the reason engineers barrel through the process to find bugs that impede the process:

> It's about encountering problems and figuring out why. It's also about attacking problems that are ill-defined and require a lot of investigative work. You can't just say you can't do it. That's an unacceptable answer. There is a reason for the problem. I can't tell you the number of times where I've been like, "this is non-deterministic,

this is total nonsense, this is impossible." It never really is. There's just a missing piece of knowledge, somewhere, and it's the wherewithal to stick it out and find it. But it's your job, so you have to figure it out, even if it makes you want to kill yourself. There is a way, it is solvable. (Informant and O'Donnell 2005a)

It is this exploratory sleuthing process—simultaneously creative, logical, and systematic—that characterizes much of the engineer's game development work, particularly during the process of preproduction. There is a kind of continual seeking for the "bottom" where the rubber treads of software systems laden with design data and artistic assets meet the road of computer system hardware. It is crucial to reiterate that this is a creative collaborative process not run by a single discipline. As is often the case in collaborative environments, tempers flare and the process begins to break down. When demands external to the development team are made (by management, for instance, or publishers and intellectual property holders) and the game development team must make decisions about what to include or not, friction is inevitable. This is discussed in more detail in World 6.

World 2-4: Instrumental Work/Play

In each of these—design, art, and engineering—there is something that drives game developers (and workers in the new economy more generally) to weather obstacles and stick with their industry. It is an aspect of the work that encourages workers to push further and harder than necessarily required. This has benefitted corporations because their workers are plugged into their work in ways they have not been previously. Perhaps on some level it is a realization of a Protestant work ethic based on the idea of a "calling."

As sociologists and ethnographers of virtual worlds (and gamers more generally) demonstrate, it goes beyond the simple answer, "because the work is fun," (which it may very well be some of the time). More than work's level of fun, there is an underlying drive—a dedication—to what Taylor (2006a) calls, "efficiency and instrumental orientation (particularly rational or goal-oriented), dynamic goal setting, a commitment to understanding the underlying game systems/structures, and technical and skill proficiency," on the part of game players (72–73). Among many game developers, the desire to know the structure of the system within which they work has led to an almost instrumental or "power gamer" approach to work/play. In part, this is a product of a system that seems arbitrary or imposed, hence the "broader ambivalence about what constitutes legitimate play[/work]" (Taylor 2006a, 72–73). The instrumental or "power gaming" of the workplace, which I have termed *instrumental work/play* evidences a desire to understand the hows and whys of the structures within which game developers move. In working so intently at having fun, artists, designers, and engineers seek knowledge that will enhance the ability of the knower in their work/play quests.

Instrumental work/play is rooted in the culture of gamers, who place significant importance on the act of working through the complex problems found in videogames. Any circumvention of this labor is often seen as a circumvention of the rules. Players are expected to play within the rules of the system, though circumvention through legitimate play is often seen as exemplary play (Consalvo 2007). Personally and deeply exploring the systems one works within is at the core of instrumental work/play. These same motivations also seem to plug into the ethic of secrecy that dominates the videogame industry. Much like "walkthroughs" are seen as the tool of the less adept videogame player, most game developers expect one another to understand the processes and practices that are, for all intents and purposes, undocumented.

Instrumental work/play spans the disciplines that constitute the category of game development. Informants often discussed how the work of making a game, either from a code, art, or design standpoint, differed fundamentally, depending on the method. Each way could be more or less difficult, time-consuming, or error-prone than other applications. The ability to understand how a complex whole, such as videogame development, is composed of numerous underlying elements (engineering, illustration, and design), contributes to the delimiting factor in the collaboration necessary for creating the final work.

Box 2.7

#: SET DEMO_MODE 1
AUTHOR_DEV_DILEMMA: Rooted in the engineer's insistence that there is a difference between "normal" programmers and the ones who thrive in game development is the argument that game programmers have a desire, drive or interest in their work that distinguishes them.
Casey: What do you mean "like you?" What is the difference between someone with a pulse and you?
ENG_DS_Spidey_1: There's a difference . . . I mean, you can get through an academic program by just going through the motions. I guess it's like working out. You can go to a martial arts class

(Continued)

```
and you can make all the motions, or you can really do it. It's
all in the energy you put in. So you can go into a computer sci-
ence curriculum and you can complete all your assignments and
you can memorize things for the test and you can walk out with a
passing grade. But that doesn't mean you know how to do anything.
Casey: Why is that do you think?
ENG_DS_Spidey_1: It's a certain . . . Depth of understanding that
you have to have. They give you all the pieces in a curriculum
but there's nothing that makes you coalesce it all into a useful
body of knowledge. I don't think that is possible without sitting
down and trying to accomplish something that wasn't assigned
to you.
AUTHOR_DEV_DILEMMA: It is both an issue of drive and initiative,
but also one of bringing things together that aren't spelled out
for you in advance. It gets at the kind of experimentation and
trial-and-error that surround the craft of game development.
#: SET DEMO_MODE 0
```

So it is not just technicians or engineers who, as historians and geographers of technology note, work at the "empirical interface between the material world and machine-generated representations of the world" (Downey 2001, 229), but nearly every game developer. Each engages with instrumentalist tendencies in an effort to accomplish their goals. Engineering and design studies researcher Louis L. Bucciarelli (1994) says that "designing is not simply a matter of trade-offs, of instrumental, rational weighing of interest against each other," that "nothing is sacred, not even performance specifications, for these too, are negotiated, changed, or even thrown out altogether" (187). But, the fact of the matter is, the design process for game developers—that push and pull—determines where the bottom is. Despite the construction of rigid specifications, games are based on something that must be felt out and determined by the players. Eventually developers run into the limits of electrons and silicon. Specifications are made, but they are not made up; they're the result of a negotiated process, which is frequently the product of instrumental play.

The trouble is, not everyone is interested in playing this way. Instrumental play has it limits. "This sense that somehow these players are just too dedicated, indeed almost bordering on the psychologically pathological, is a popular theme. What I found in conversation with power gamers, however, is that they consider their own play style quite reasonable, rational, and pleasurable" (Taylor 2006a, 72–73). If, as Taylor notes, "these functions are mainstreaming the focus on quantification in that play style," then likely this work/play style will become more widespread. "But if the adoption of these tools and with it the play styles it brings becomes mandatory," Taylor (2006b) asks, "must we start to deal more concretely with notions of emergent coercive systems?" (332). This seems to get more directly at the heart of instrumental work/play.

I also like to think of instrumental play as an alternative narrative to the dominant discourse surrounding "casual" versus "hard core" gamers. Many game developers no longer count as "hard core" within this dichotomy; nor are they casual, either. Instead I posit another term, *instrumental players*, who are dedicated not to a particular genre or subset of games, but who consistently and persistently attempt to dissect their games from the mechanics up. Many designers play very few hardcore titles because the mechanics are instantly recognizable, and their interest lies in uncovering structures. Many so-called casual gamers are adept instrumental players. Their ability to strategically change their play based on knowledge of the underlying system is precisely the kind of instrumental rationality and sensitivity for the underlying game mechanics that are so crucial for game developers. However, many instrumental gamers find it difficult to simply observe or play games because they have difficulty resisting the urge to determine how a system functions. They continually see the underlying systems and may find it difficult to participate in either casual or hardcore attitudes, immersing themselves in a particularly complex game intensely until they feel adequately satisfied that they understand the underlying systems that make it function.

Instrumental play should be distinguished from a kind of "instrumental rationality" or "instrumental reason" as it might be defined by critical theorists of the Frankfurt School (Adorno and Horkheimer 1976). Instrumental play is distinguished from these theoretical categories in that it has no claim to the irreducible or absolute. In fact, instrumental play would continue to probe into the structures of what is considered irreducible. As my informants might put it, "You've got to get over that [a commitment to the absolute] real quick."

Instrumental play is about searching out associations, analogies, and relationships, much like "enlightened" scientific inquiry, but it makes no assumptions about the absolute character of those suppositions. This is where the "play" component of instrumental play is crucial. There is always the assumption that what you are working on or working with will swerve and send you in new directions. This is more in line with the idea of the game developer as "bricoleur" (Lévi-Strauss 1962, 17), adept at performing numerous diverse tasks, "mak[ing] do with whatever is at hand"

(Lévi-Strauss 1962, 17). The concept of bricolage, or the bricoleur, is not new to the studies of technological development, but instrumental work/play plugs into the bricoleur's underlying drive, which is to push one's tools to the brink and pull off "risky" moves, doing what others have thus been unable to do. Put another way, instrumental work/play is what pushes bricoleurs to attempt creations that strain their understandings, no matter the extent to which that bricolage understanding appears to be "reality." There is always a time and place to question the bricolage system that one has constructed in an effort to pull off a new feat of creative work.

Unfortunately, in many cases developers do not even have the opportunity to form specifications or play in instrumental ways, methods that are crucial to pushing the tools, expertise, and products of game studios. Developers doing outsourcing work are frequently given rigid guidelines for their work. Unlike developers working in studios, outsourced work is rejected outright if it does not meet specifications. Developers' ability to get at the underlying system in need of changes has been compromised because there is no link between the specifications and the process from which they were derived. The ability for an artist or designer to see their work within a game is obscured when the work is done by contract developers since, often, only the contracting studio has the ability to view the results. Anthropology of work shows that in the "absence of informal working knowledge from technologically sophisticated production processes. . . . technical workers. . . . develop complex cognitive models that represent the . . . messy world of work" (Baba 2003, 19). While these models provide the frameworks that other workers can use to get things done, developers remain hamstrung by their inability to access the underlying systems from which these demands arise.

For many developers, the ire at being pushed into situations where premature designs are released or key elements are cut is directed at management. Studio heads and managers play with their organizations like artists play with textures and models, but for ends separate from the game. Studio heads are often myopically focused on either the continual survival of the company, or its acquisition by some parent organization. They must interface and play with game development studios, intellectual property holders ranging from movie studios to comic book publishers, console videogame system manufacturers, and videogame publishing companies. Studio heads and managers also bump into the limits of their teams, their employees, their networks (both social and technological), and their access to secrecy networks. Managers must do as much as possible with as little as possible. Teams will get shuffled around based on the work available. If particular designers have proven themselves able to handle particularly restrictive conditions, they may be moved from one project to another in the hopes of bringing new perspectives to existing teams. Just like their subordinates, managers must often instrumentally play within the structures of a project that they have been hired to complete, without necessarily having any say in the parameters by which they must abide. Creative visions may differ, and those who do not need to rework systems to achieve modifications seem to ask for constant modifications. Representatives of corporate institutions use a game that is in development as a means of negotiating with other institutions. Studio managers must also frequently engage with the management of a system, which in many cases began as play, for many studio heads began as developers working on a game in one form or another.

Time and again in my fieldwork, the interface between the creative, logical, systematic, and intuitive met with the emotional. At times it was due to conditions outside of the control of my informants. The demand for overtime or crunch, an occurrence examined more closely in World 5, would often cause collaborative practice to break down. Perhaps most disturbing were those occurrences where particularly unwavering restrictions about how a system should be implemented were brought into a project that had already progressed in a very different direction. In these cases, developers routinely deflated and proclaimed that "this would be so much easier with system X," or grew frustrated because they knew that "this is so much easier in program Y." This emotional response coupled with a reversely proportional emotional response on the part of developers already part of the project proved explosive. Thus, the phrase "you gotta get over that real quick" rang through my head time and again as I watched communication practices break down among developers.

The developers' tendency to force themselves beyond that which the job requires does not stem from a demand or coercion that everyone push harder and play/work in the same instrumental ways. Overt coercion undeniably happens, but the structures that many workers trapped in the new economy must work within are also key constituents of why people feel compelled to work and play in instrumental ways. It is not simply the systemization and regimentation of game/sport that causes the loss of game-like innocence. The issue is larger than that. And the argument is not simply a "game for sport versus game for money" dichotomy. Simply bringing money into the whirlwind does not automatically cannibalize your game. The difficulty is that money brings those interested in playing other kinds of games—often exclusively financial games—into a world that many had hoped might stay a game. It is the incorporation of a drive toward institutionalization that changes the play. As anthropologist, John Kelly has written about American baseball:

> But it wasn't commoditization that changed baseball so unmistakably. It was higher levels of capitalist organization. Above all, the leagues changed everything. What are they, and what is their relationship to commodities? Commoditization, yes, but we will need more tools than that: we will need to understand whole new layers of management. We will need a theory of the firm. Professional Sports leagues did more than commoditize the game. They incorporated it. (Kelly 2006, 55–56)

So what has changed the play of game development into the work/play of game development is the coalescing of a willingness to play (or be coerced into play in particular ways) with the systematic incorporation of the videogame industry. That move to industry rather than something else marks an event that begins to alter the space of play. While baseball is one example, the connections with the videogame industry are undeniably industrial. Games do not start as an industry, but have moved into that space. It is not simply that people good at playing a game begin to accept compensation to publicly perform their play. That might make it sport, but that alone does not account for the work/play conflation.

Kelly notes that bringing money into baseball opened the door to monetary concepts like free enterprise and profit, the result of which was a prioritization of interested parties over the game's interests.

> Competent professional players began to make the game a means to profit, starting with the prowess of the Cincinnati Red Stockings. The line of movement from clubs to leagues to Organized Baseball remade baseball into an increasingly interconnected congeries of commercial institutions, made baseball into a branch of what Americans like to call "free enterprise." Baseball reorganized from independent clubs, originally player-oriented leisure groups, into profit seeking corporations in a legally powerful cartel. We can use this history to discuss dynamics of capital, profit, and finance in the actual capitalist world. What are the key genres of capitalism, its defining institutional structures, drive belts of its history? What is baseball, when it is not only a genre of game, but also, a genre of capitalist enterprise? What then constitutes its best interest, and how? (Kelly 2006, 59)

Baseball and game development are each a genre of game work/play. It is the connection with commercial, profit-driving organizations that has so dramatically shaped game developers' worlds. Profit also provides us a normative point of entry. What drives this bus and how does it go? More important, "what constitutes [the game's] best interest, and how?" In many respects, it may be a return to these questions that has signaled a resurgence of experimental, artistic, and independent game development. These movements are attempts to recover a kind of lost history rooted in games being about a love for finding and playing with underlying systems, which was supplanted with a drive for systems that encourage profit seeking.

There is a reason that games now find themselves at the forefront of a wide variety of economic frontiers, ranging from the gameification/ exploitification/bullshit debates (Bogost 2011; Kazemi 2011; Zichermann 2011) to continual acquisition and shutdown of small game studios. This largely deregulated or self-regulated space exists at the boundaries, where rules have yet to be constructed or even discussed and thus the decision making is often left to those with the largest bank accounts. Though game development may seem distant from other frontiers where extraction results in death, cancer, and destruction, it too suffers from being at that edge where the "expansive nature of extraction comes into its own" (Tsing 2005, 27). These new debates about what games are, can be, or ought to be are the questioning of the importation of new frontier profit seeking ethics.

World 2 Boss Fight: "Ya Gotta Get over That Real Quick . . ."

Preproduction is only successful when it begins to orchestrate the collaboration between the teams working on a project. In my research it was common to see projects that began with the best intentions, but failed because a team was unable to construct the necessary tools for collaboration as a project moved into production. One such example was a recently hired Artificial Intelligence (AI) programmer assigned to lead a short-timeline project aimed at bringing an existing game "franchise" (referred to here as *Lure of Action*) to the PSP. This particular engineer was exceptionally knowledgeable in both AI and game engine development, but his ideas about how the game should be implemented were very different from those throughout the rest of the company. The project team, which was in preproduction at the time, took faith in their lead engineer and made the departure from the path of the rest of the company. However, as production loomed and their pipeline was not ready, the team glimpsed the significant ramifications of their divergence. None of the other tools and software systems developed throughout the rest of the company were compatible with these new approaches, and most of the other engineers within the divergent team were unfamiliar with the approach taken by the new engineer. Thus, the emotional decision to go with one engineer's instinct became a serious liability for the project team.

This example is indicative of the kinds of breakdowns that occur commonly in the process of game development. At the core of creative collaborative practice is the ability and necessity of being able to play with and get at underlying systems—technical, conceptual, and social. When access to underlying systems is undermined, so too is creative collaborative practice. In the rogue engineer example, extensive disconnects rendered one piece of the system useless because the engineer refused to determine if there was some other mechanism by which to reconcile the approaches of the rest of the studio.

This steadfast approach, when it ignores the circumstances, evidences another emotional response that hijacks healthy creative collaboration. There is often a kind of bi-directional lack of respect for differing approaches to development within game companies. Many experienced developers scoff at the ideas of inexperienced, young developers. At the same time, young game developers will quietly mock the approaches taken by senior developers. In only the best situations are actual bridges made across these divides where developers explain to each other why a particular approach might be difficult or problematic later in a project or why something overlooked might offer a solution. Stubborn refusal to change their ways, emotional assumptions that nobody else could be right, and lack of productive communication are developers' foibles that often derail projects.

The same can be said of the lack of acknowledgment that game development requires significant interdisciplinarity and skills variety. Engineers will often privilege engineering labor, designers privilege design elements, artists privilege artistic efforts; other disciplines such as sounds, writing, and production will claim to be the key(s) to game development. But arguing about importance ignores the fact that any single element on its own would be something quite different and would do something quite different from a videogame. It is the coming together of all these elements in a way that works as a system of systems to convey one or many concepts that makes videogames functional and interesting.

Interdisciplinarity is at the core of videogame development, and as such it needs to be presented this way, not merely as a world dominated by engineers. The broader fantasy about what game development work is and who game developers are shapes who becomes interested in making games, and thus the kinds of games that are made. Interdisciplinarity is tempered by the ability to communicate and work across disciplinary divides, processes aided by the creation of new categories of specialization: tools engineers and technical artists (further discussed in World 3-3). This specialization can be difficult to achieve in young or small studios just starting to explore the work of game development, but the importance of these new categories of specialization is rooted in experience.

As game companies grow they rediscover the importance of developer specialization, yet more broadly, the industry does not communicate or express this integral component of game development work. Despite rosy collaborative pictures like the one painted below, interdisciplinary work takes time and the emotional maturity to accept that the ideas brought by each area of expertise are worth considering. If any one component of the collaborative team is unwilling to recognize this, the system breaks down. The project manager on *Resistance: Fall of Man* commented on the importance of interdisciplinary collaborative practice and how it was something that the studio, Insomniac, had to continually work at during the more than two-year development cycle.

> Insomniac grew from a company of 40 people to around 160 in a few short years. In order to keep the business running smoothly, a new layer of management structure was introduced, which worked surprisingly well. But Insomniac quickly became more departmentalized. People began to focus more on the needs of their department than how their department related to the ultimate goal: the game. . . . By the end of the project, it was common to see animators sitting next to gameplay programmers, going over get-hit timings and whatnot. In a collaborative environment where each person brings ideas for improvement and innovation, getting the right people together is the key to creating quality. (Smith 2007, 36)

Many game development studios remain departmentalized, which isolates collaborative resources within disciplines. Both interdisciplinary and intradisciplinary collaboration is productive given different circumstances. The key is finding the time to pursue collaboration cautiously. Interactivity can be measured and tentative; studios do not need to jump blindly into interdisciplinary collaboration.

Because hesitation is both warranted and compatible with implementing interactivity, this world's boss fight really amounts to a quite simple rant: "you aren't going to do this on your own." It takes many people to make games and it is because of developers' ability to see game worlds in different ways that the entire production process works. Certainly it is possible for an individual developer to create small experimental or artistic games that work and are quite compelling. But these are not the games that reach large numbers of players. They may grow in popularity over time, but this is typically only after a coalescing of other perspectives.

The interconnectedness of game development work cuts to the heart of what makes it able to produce the technologies it does, as well as why it can be so unpredictable and complex for those working in it and those attempting to manage the globalization of this industry. Game developers stress not only the sheer number of disciplines that work on the creation of games, but on the numerous forms of communication that ultimately becomes a goal in its own right. Despite this, developers continue to talk past one another. The relationship between interdisciplinary work and communicative collaborative practices are constantly on the minds of developers.

Furthermore, too often, while a game is credited to the numerous developers found in a game's actual credits screen, it is cited as the work of an individual; a producer or designer. Yet this neglects the fact that the entire process of preproduction is about the coming together of people possessing different expertise and capacity to think about what a game might be, how it might come to be created and begin that process. Game development is about interdisciplinarity, thus any reductionism with regard to what/who a "developer" is neglects those other disciplines that make games possible. This isn't about a single individual or one discipline guiding the process. It is about the assembly of a space where creative collaboration can occur. Any commitment to a single person's ego, approach, or perspective will only end in disaster. Ya gotta get over that, real quick.

World 3: Assembling Experimental Systems

Box 3.1

#: SET DEMO_MODE 1
AUTHOR_DEV_DILEMMA: The pipeline is one of the most important things being assembled during the preproduction phase of game development. It is the set of practices, tools, standards, aesthetics and themes that will carry the project into the production phase of game development. It involves defining a rigid enough set of specifications and systems so that the game will hold together, and maintaining a flexible enough structure to allow the production team of the game development studio the ability to be creative within those confines:
Casey: The tools and pipelines have worked well for you?
ART_DS_Ogre_1: I was expecting there to be a more organized pipeline I guess. [Laughs to self] I always think of *Super Mario Bros*. when I hear the term "pipeline." I just imagine going down it.
AUTHOR_DEV_DILEMMA: For some artists, the pipeline is the set of tools made available for production. It is a set of systems and practices that ought to already be in place. For others it is the very aspect of game development that interests them. For those interested in creating the pipelines that will be used by others, many find themselves toiling on the assembly line in submission to those systems:
TECH_ART_1: Production . . . Well, it is a drain on the soul. I was the lead character artist for COMPANY_X. I made like 150 characters for this game. Functionally they were all exactly the same; they just looked different. I might be better with production as a designer and getting to make game play that acts different for different parts of the game, but what I was doing was window-dressing. That is just, "go make this pile of stuff so people can perceive it as being all the content."
Casey: So, preproduction is your favorite?

(*Continued*)

```
TECH_ART_1: Well, not preproduction specifically, but it is solv-
ing problems and making decisions about the big picture about
what the game is going to be and how it is going to be. I would
much rather figure out how this really complicated character
model is going to work than go and make ten of them. I have done
both.
AUTHOR_DEV_DILEMMA: Yet, the goal of preproduction, is always to
create a set of technologies and practices that will ultimately
be put into production. Put another way, at one point or another,
game development will turn into the assembly line process where
materials are being produced for the game at a rate that will
ultimately strain the very systems designed to accommodate them.
#: SET DEMO_MODE 0
```

World 3-1: The Importance of "the Pipeline"

The pipeline, for videogame developers, is the set of technologies, standards, and practices through which art assets and design data flow into the underlying game code. It is processed and ultimately displayed on the screen as a game. Pipelines are often highly specific to each project because there may be particular aspects of the pipeline that are used for one project, but not for another. The pipeline is simultaneously a thing sui generis and motley. It would be difficult to go into any game company and ask to "see their pipeline" (and it would be unlikely that they would show it to you without having signed an NDA). Even if you gained access to the pipeline, you would only get glimpses and would not get to see the whole thing.

The pipeline is one aspect of game development least talked about by those outside of the game industry, probably because it is something to which only experienced game developers have paid particular attention. Game development hopefuls rarely encounter a detailed description of how "you get a bunch of stuff" into a game after you have conceptualized it. Most developers never get that far. The pipeline rapidly recedes into the background of game development as it is constructed. In some cases, pipelines may simply be a wiki page(s) on a web server that describes how to properly tag models or animations for export from *Max* or *Maya*. In other cases it will be a wiki page(s) combined with a specialized button placed in *Max* or *Maya* that automates the process. However, many of those things quickly become taken for granted as a project proceeds from preproduction to production. For our purposes, the pipeline and all its requisite practices, technologies, and standards need close attention so we can discern how pipelines come to be developed.

Pipelines and tools shape the possibilities of game developers in profound ways. All of these technologies, standards, and practices combine in the "experimental systems" that developers create to enable the development of videogames. The desire to tinker and have highly responsive feedback loops between one's work and the final game results in a tightening relationship between worker and product—only if the systems enable such tinkering. Throughout the initial stages of preproduction, artists, designers, and engineers feel out limits, unearth fundamental structures, and put in place underlying systems. The next step is the creation of tools and pipeline that enable continued play on the part of developers as they continue down the road of game development.

To begin to understand the pipeline, we must elucidate the focus of developers' work: essentially they work to enable the work itself. Flexible technologies, built on systems like XML or scripting languages like Python or Lua, have become the new key component of game development. They offer the ability for artists and designers to alter a game's characteristics, such as the power of a weapon or allowable configurations of a player's gear. Engineering must frequently create these tools that are at the core of what makes game development practice function. The lead engineer on *Battle Engine Aquila*, a game that was developed for the PlayStation 2 and Xbox over the course of two and a half years, notes the importance of flexible and modifiable systems.

> Flexible core technologies [went right]. As much information as possible was read in from externally editable files, and several custom editors for different areas of the game were written to allow designers and artists to alter everything from level layouts and unit statistics to graphical effects, without needing code changes. . . . This approach paid off both by reducing the knock-on effects of changes and potential bugs and by enabling a lot of experimentation during the game's development. (Carter 2003, 51)

At their best, flexible technologies provide members of a game development team the ability to work independently. The ability to experiment with ideas in the stream of constantly shifting sets of properties, enables each member to individually push the boundaries of creativity. In the end, the use of flexible technologies becomes good design practice; the ability to expand or contract game components without the cumbersome team intervention allows each developer to experiment simultaneously.

These flexible technologies also interface in game development studios with the numerous disciplines that birth them—artists' tools like Photoshop or 3D Studio Max must interface with data created by designers in the form of scripts or XML, for example. The process, therefore, demands engineers who work well with designers and artists; otherwise you will end up with mutant technologies that merely reinforce the old ways rather than bridging artists' and designers' understanding the world. One of the most critically acclaimed games of 2007, *BioShock* (which was released on the PC and Xbox 360 and took nearly three years and four thousand files to create), suffered difficulties, as noted by the project lead: "Many of the processes and tools we used to develop *Bioshock* were inefficient or confusing in implementation, leading to slow iteration cycles and bugs" (Finley 2007, 26).

Tools enabling work interaction are frequently written in the "spare" time of engineers, when other demands are not being placed on them to provide basic game functionality. These tools, if poorly designed, become hazardous to the health of a project. Wedging flexible-technology building into the small cracks in engineers' time forces the development team to accept something that might not work. When their play ends up failing, many designers and artists are convinced that it is their fault, rather than approaching an engineer to understand why a tool or process is not working.

The mechanisms that enable developers to interact with their systems, data and each other are infrequently discussed or shared. Secrecy between teams is the modus operandi throughout game development. Even if flexible technologies are shared between teams, critical information, such as the social practices that surround those tools are omitted or are simply referred to as "tools." No explanation is given as to what these technologies do or what they accomplish for game developers, nor how the designers, artists and engineers can improve their work using flexible technologies to work together or independently. These tools are unknown until someone has begun working in the game industry, and are generally cloaked within each team even though these tools are cited as one of the most important components of the game development process. The supervisor on the game *Final Fantasy XII* for the studio and publisher Square Enix notes the importance of tools that provide experimental or trial-and-error approaches to design: "Our various in-house authoring tools, coupled with commercial digital content creation tools . . . created an environment in which we could use trial-and-error tactics with the new tools while also increasing productivity by using the ones we already knew well. It was especially helpful for us that the in-house tools enabled real-time previews using the game's rendering engine (Murata 2007, 24).

The existence and development of tools that allow experimentation and heighten productivity should be touted throughout the industry but flexible technologies remain a game development industry secret, largely unknown and unexplored by those looking to enter the videogame industry. The explosion of developer excitement surrounding development tools like *Unity 3D* are due almost entirely to how these software systems illuminate the previously obscured territory of flexible technologies. The methods and practices used to actively produce games have remained the "black art" of game development. Technology has not held developers back nearly as much as obsessive secret keeping and naïve neglect of practice have.

More disturbing is that these flexible technology workflow resources are most frequently kept from companies doing offshore outsourcing work for videogame studios. "Real-time previews" of game's content inside of a "game's rendering engine" is frequently cited as essential by artists. Yet time and again in India I encountered artists struggling to work within the confines of structures unknown and invisible to them because the experimental tools created by engineers specifically for the project, which would enable these contract developers to understand where, how, and why aspects of their work were failing, were withheld by the contracting organization. These artists were not able to preview any of their work and therefore spend considerable more time and energy frustrated, unable to explore options and blindly searching for answers. As I observed them, they would attempt to respond to annotated renderings of a model as seen in engine, though they were unable to tinker with the model and immediately view it in the game's engine. Changes had to be filtered through contacts at the contracting company often with significant time lag. Presumably, all the contracting company would have to do is run the models through the very preview tool the Indian developers so desperately needed. The gate-keeping model fails on so many levels, not the least of which is the time and money lost while developers sit around and wait, unable to work. Of course if at any point those tools require a component limited by licensing or NDA, the ability to share those tools further breaks down.

Processing everything by hand the way contract developers are forced to is the "old way" of developing games. This approach requires an artist (or a designer) to harass/bug/pester someone (frequently an engineer) so they can see their work operating within the game. Most game developers work this way early in their careers for lack of knowledge of the independence and efficiency of flexible technologies. The lead designer of *Diablo II* (which was created by Blizzard Entertainment, one of the largest and most respected game studios among developers) notes the difficulties that result from not having these workflow tools created. *Diablo II* took more than three years to develop and required a twelve-month crunch period.

> We developed the original *Diablo* with almost no proprietary tools at all. We cut out all the background tiles by hand and used commercial software to process the character art. . . . The greatest deficiency of our tools was that they did not operate within our game engine. We could not preview how monsters would look in the environments they would inhabit. We couldn't even watch them move around until a programmer took the time to implement an A.I. Even after that, an artist would have to hassle someone to get a current working build of the game to see his creation in action. . . . Our lack of tools created long turnaround times, where artists would end up having to re-animate monsters or make missing background tiles months after the initial work was completed. (Schaefer 2003, 88–89)

The producer for *Crackdown* on the Xbox 360, a project that took nearly four years, also notes the significant lag times created by poor experimental tools: "The testing of a single asset could take upward of an hour, directly impacting productivity and indirectly impacting quality since it naturally discouraged regular testing" (Wilson 2007, 30).

More than discouraging regular testing, developing games "the old way" discourages exploration and experimentation, which is crucial for developers to push the quality of a game forward. Those companies that tout the sharing of information, tools, and practices continually cite flexible technologies and tools as the key that took the game from mediocre to excellent. Yet for many developers, this painfully slow and less flexible process remains typical and results directly from the unwillingness of the industry to share tools and processes. This reluctance to share or improve does not provide the kind of experimental environment that seems necessary for the creation of videogames. Aspiring game developers and newly created videogame companies continue to function without the tools they need and without the ability to do their best work. For developers in India and young developers in the US, there is no "new way" until experience and access allows it. Developers cannot do their best work, as they are continually asked to do more, work more for the same cost, often with similar or worse results. The issue is not one of productivity, but rather sensibility.

New mechanisms are necessary to increase the productivity, creativity, and efficiency of work. And perhaps more important, the sharing of practices and tools surrounding pipeline development is crucial for an industry struggling to deal with quality of life issues. Both experienced and inexperienced teams would likely benefit from decreased secrecy surrounding aspects of game development that have proven useful, since the more flexible technology and key tools can streamline development work, the more developers are free to experiment and to create better games. Yet, the developers often lack the proper tools, and the process therefore breaks down. "Production difficulties," is often a code word for breakdowns in the carefully constructed pipelines and timelines built during preproduction. One developer, in a discussion about some of those difficulties even marked problems as still "in preproduction":

Casey: You were talking about production difficulties on *SM3* . . .
DESIGN_LEAD_1: Yeah . . . Well, preproduction on *SM3* actually went fairly well, all things considered. I take that back. Preproduction for design and art went really well. Engineering preproduction lasted almost the entire process.
Casey: There were pieces coming together fairly late I recall.
DESIGN_LEAD_1: Yes. The technology really fell apart there, but the other efforts were managed well and went fairly smoothly. (Informant and O'Donnell 2007)

In some cases, when a pipeline or tool chain is effective enough, it becomes a company asset, such as the Unreal Engine and its host of tools, which were developed originally during the creation of the first *Unreal Tournament* game. Other technologies such as Unity 3D are designed from the ground up in real time as developers run into pipeline issues such as asset importing and data management. There are numerous tools such as these that, to one degree or another, aim to streamline the game development process by automating the pipeline and its capabilities. Yet, these tools can be expensive, may not support the platforms needed by developers or have capabilities locked until separate licensing for a platform has been acquired. These development platforms, in many cases, also presume certain types of games and can prove difficult to use in other cases. Game development is often a labyrinthian process, as designers, engineers, and artists muddle their way toward a game. As noted in World 2, the experimental character of the work often means the requirements for any given tool will be pushed to their breaking point during the process. In other cases tools will need to be changed out in favor of others as a project moves forward or at the demand by parent company or licenser. Unity 3D, in particular, has caused a significant shift in the game development ecosystem. Both experienced and amateur game developers now have a tool available for developing games that enables a wide variety of game types and experimentalism on the part of developers. At the same time, it is possible that Unity 3D may continue to be viewed by some as a tool for amateurs who have not yet transitioned to "real" game platforms.

I use the metaphor of the labyrinth and experimental system in this world because it connects our enjoyment of working within limits to pushing those limits. "Not anything goes. If there is construction, it is constrained." Game developers and scientists alike "meet with resistance, resilience, [and] recalcitrance" (Rheinberger 1997, 225). Put another way, the tools of game developers are shaped by and shape the kinds of games they attempt to create.

As previously noted in World 2, engineers, artists, designers, management, and corporate executives all find themselves exploring the fine interface between work/play. Engineers will interact with and attempt to better understand the underlying hardware systems that structure their code. Artists will adjust models in ways that work within rigid confines defined by engineers, attempting to create aesthetic harmony in a highly structured space. Designers will adjust maps, levels, and scenarios to bring out particular player experiences. Management teams will move around team members in the hopes of maximizing employee output without derailing teams or projects. Executives will change release dates in the hopes of better positioning their offerings against competitors' offerings or to coincide with marketing pushes made by console manufacturers. This constant state of flux that rules game development practice further suggests the importance of work/play for developers. Thus, these tools and software systems enable the kind of experimentation and play so critical to game development.

And so, the developers who actually gain access to a pipeline come to see it as a kind of savior for the game development process. Because the pipeline is always co-constructed during the process of preproduction, however, it cannot completely guard against the inherent unpredictability of the process. As historian of biology Hans-Jörg Rheinberger writes about experimental practice and experimental systems:

> An experimental system can readily be compared to a labyrinth, whose walls, in the course of being erected, in one and the same movement, blind and guide the experimenter. In the step-by-step construction of a labyrinth, the existing walls limit and orient the direction of the walls to be added. A labyrinth that deserves the name is not planned and thus cannot be conquered by following a plan. It forces us to move around by means and by virtue of checking out, of groping, of tâtonnement. He who enters a labyrinth and does not forget to carry a thread along with him, can always get back. (Rheinberger 1997, 74–75)

Given the complexity and frequent demands for secrecy, it is not surprising that developers thus far have been constructing the labyrinth of game development as they go. It is no wonder that in the rush forward they have not bothered carrying any thread along. Developers have typically made a headlong plunge in, with no way to get back, or even untangle where they have been. They occasionally make retrospective assessments, "postmortems," wherein they discuss where think they have been and where they might have gone wrong. But postmortems are not successful enough to keep development teams from continually take the same wrong turns.

The situation becomes more intense when developers realize that they must maintain the secret society of the office and industry. Suddenly they are not willing or able to talk about the labyrinth in any real detail. They talk about how pretty the vines look, or how they were able to grow them in a particular way, or that sometimes you take wrong turns and have to work late to find your way back. But postmortems are often internal discussions, applied sparingly to future projects and never to interstudio understandings. Because the construction of each labyrinth is seen as unique and developers discuss the full construction practice so rarely, it is difficult to connect practices throughout the industry. Meaningful collaboration at the industry level has been rendered impossible by the culture of secrecy that plagues the game industry.

The implications of carefully guarding secrets about lessons learned and labyrinths built, as seen more broadly throughout businesses in the new economy, is troubling. If game development is an index into new economy work—both are spaces where experimental practice is crucial—then the ability to communicate and think about access or obstacles to experimental practices are even more important. In game development, demands for secrecy seem to have taken precedence over the maturation of game development practice and, by extension, threaten the development of the new economy.

In a community where play and games hold such high meaning, secrecy is even more insidious than a simple mandated code of silence. There is a game in secrecy as well. "Tension means uncertainty, chanciness; a striving to decide the issue and so end it. The player wants something to 'go,' or to 'come off,'" we want to succeed by our exertions (Huizinga 1971, 10–11). Again, the metaphor of the labyrinth in experimental systems ties to our enjoyment of working within limits, but also to having those limits pushed.

But how do we extrapolate from videogame systems and structures to experimental systems and the interactivity of people? Experimental systems have become a useful way to think about game development, in particular the work of designers, those people who end up interfacing with the work of engineers and artists. Their tools are created by the tools engineers, but frequently with a mind toward changes down the road. This is also why you have tools engineers and technical artists accompanying our new systems, technologies, and practices. As sociologists of science have shown, "the more automatic and the blacker the box is, the more it has to be accompanied by people" (Latour 1987, 137). In part it is because these "outcomes are often not consciously calculated, or even intended by any one of the parties involved" (Knorr-Cetina 1983, 130). Because they are embedded in a broader social context of practice, they must somehow retain those connections.

> Experimental systems are to be seen as the smallest integral working units of research. As such, they are systems of manipulation designed to give unknown answers to questions that the experimenters themselves are not yet able clearly to ask . . . They are not simply experimental devices that generate answers; experimental systems are vehicles for materializing questions. They inextricably cogenerate the phenomena or material entities and the concepts they come to embody. Practices and concepts thus "come packaged together." (Rheinberger 1997, 28)

Until a project reaches production, and frequently even after that, almost every aspect of the game development process must act like an experimental system. It must be open, or capable of providing unknown answers. Sometimes these unknown answers are frustrating, but often they become aspects of the game proper. By necessity, experimental systems must be interactive. They must respond in real time to other technical systems, data, and people. This is where the headlong rush forward begins to be a dangerous practice for game developers. As designers play with all of the art and code assembled with experimental tools being used by multiple people, interactive technologies obscure the traces of what happened where. Multiple experimenters running multiple experiments render even the remotest possibility of an accurate reconstruction of the past next to impossible, even for game developers. Rather, many developers and engineers hope for the possibility of "mind melding," because experimental systems and interactive development tools make it impossible to map where the game came from and where it will eventually be when it becomes a golden master (GM).[1]

The reality of relatively secret and exploration-obscuring experimental systems, then, is about change. Like a labyrinth itself, game development practice has built and is constrained by its histories and technological systems. Because of this, many experienced developers rightly insist on being able to manipulate things on the fly, interactively. Developers bump into their histories and technologies, and in the cases where every exploration requires going back to the drawing board, it takes years of frustration to get games accomplished. Tools including the pipeline could be the answer.

However, the pipeline is not created in a day. The emergence of the pipeline is part and parcel of numerous other changes that have emerged in the game industry. The remainder of this world examines the emergence of two subdisciplines embodying ideal specializations in the pipeline: the technical artist and the tools engineer. This text contextualizes their rise to being by examining the emergence of technological mechanisms that predate the development of new disciplines and new technologies.

World 3-2: The Console and the Debug Menu

If the pipeline and tools emerged to meet the experimental needs of game developers, then the "console" (not to be confused with "game consoles") and the "debug menu" are the historical roots of these practices. For many developers they remain the primary means by which developers can tweak a game's systems. The experimental tools that game developers create are an appeal to the kinds of interactive systems that they design for game players to interact with their creations. These systems are designed in ways that appeal to the very form that they instantiate. Game mechanics can be "played," or "tweaked," into "balance" via many of these tools. The "debug" menu or "console" found in many games seems to be one of the foundational means by which developers have attempted to make their tools more flexible. The game must act as both finished product and experimental system for game developers. It must be constructed in a way that is both flexible enough to accommodate change, yet rigid enough to ensure that it is used as it was intended.

The console is a powerful tool, but might be the least intuitive and least interactive of the experimental systems created by developers. It is the videogame equivalent to the MS-DOS prompt in Windows or the Terminal in Mac OS X. The console is commonly used to issue specific commands to the underlying game system, commands that must be memorized or have references provided for them. For example, variables can be assigned new values, thus adjusting the underlying mechanics of a game system. Boolean variables can be turned off or on, indicating whether or not particular program paths or options will be executed or optimizations made (or not).

In many cases consoles are the interface into the underlying scripting engines that have been created as interfaces designers can use to manipulate the game spaces presented to players. Scripts can be typed directly into the console (or new functionality temporarily defined for testing purposes), yet working with the console isn't an easy a task. The developer issues commands then returns to the game to see if or how those changes affect the overall game system. In some cases the changes may only be temporary, being reinitialized each time the game is run.

In some games, the console actually remains visible to the player, or just outside of view for the player. One of the distinguishing characteristics of the console, however, is the requirement of a keyboard for use. When games are being developed for videogame consoles, like the Nintendo Wii, Sony PlayStation, Microsoft Xbox, a keyboard will generally not be present for the player. In other cases, developers do not wish to expose the kind of flexibility provided by these tools to the actual users of their games. In this case, the console may only be available on an internal build of a videogame that runs strictly on the computers of developers. But when the console does remain as a feature for consumers, it functions for players the same way it does for developers, allowing interactivity and flexibility to those who know how to use it.

In many first-person shooter games, pressing the "`"/"~" key on the keyboard will cause the console to slide down from the top or bottom of the screen, providing a location from which commands can be issued. New maps can be loaded, settings adjusted, or statistics gathered. In most cases, the console is a tool for game development teams during the creation of game and is usually stripped from shipping games. But in some cases the console remains intact, especially if user-created "modifications" (MODs) are an important aspect of a game's appeal. The console provides necessary feedback to MOD developers and provides the interface through which they will often load or unload their changes to the underlying game system.

The MODs and the console are an interface between users and developers. Perhaps, even more than an interface, these tools are the means by which some users become interested in game development (Postigo 2007; Taylor 2006b). Some users delight in experimenting with these resources and in turn become aspiring game developers. These tools provide a kind of shared opportunity for developers and users to explore the underlying systems of a videogame's systems.

In addition to being a development tool for developers in specific games, consoles are somewhat common in large software systems, especially when software packages offer a large variety of functionality to their users. Consoles and scripting systems allow for the automation of particular processes or customization of a program's behavior to better suit the users at a particular site. Software packages like 3D Studio Max and Maya are two game-industry-specific examples of software packages that expose a great deal of functionality and customizability through their consoles and scripting systems. This will be examined further in the section on technical artists in this world.

The debug menu, on the other hand, while quite similar to the console, will often offer the developer a range of predefined options. Exposed variables will be listed, with current values and the option to change them. In some cases, specific actions can be executed: characters can be spawned or destroyed; models, textures, sounds, and other options can be substituted; levels can be loaded, missions launched, or specific cut scenes played. But, most important, the debug menu visually offers developers a range of options. They need not necessarily know the commands that make a particular action occur, and it therefore makes changes and exploration easier for the development team. The menu provides the user with the information regarding what can be changed and provides context. Rather than commands and syntax needing to be understood in advance, the menu presents those options and ensures "proper" use of the underlying systems.

The debug menu is nearly always removed from the final releases of a game. It is stripped from the game released to players, as its target audience was the developers making the game and not those playing it. Early in the preproduction of *SM3*, I noticed engineers and designers both working to define debug menus that enabled personal experimentation within the emerging game. Many options from within the debug menu may very well allow situations or events that were eventually removed from the final game. For the debug menu to remain in the game would create the possibility for events or possibilities that developers do not wish to occur within the final version of the game. The debug menu represents precisely the kinds of experimental tools crucial for pushing games in new directions. It provides a framework and context for understanding, but it also creates the possibility for excess, to push the underlying systems and structures outside of their intended boundaries. At the same time, this is precisely what makes it a useful tool for game developers. With little fear of catastrophic failure, designers, artists, and engineers will experiment with new ideas or approaches. This is a key feature to what makes experimental systems productive apparatuses for scientists and game developers alike.

Both consoles and debug menus evidence a deeper shift happening at the core of game systems. These tools are simply a first step down a path of moving design data outside of the core source code of videogame systems, a necessary step, as we have seen, to facilitate the kind of unfettered access needed by designers and artists. The console and debug menu are the most obvious form of a broader movement within game design and development. "Data driven" design is nothing new in the lives of software developers, but for many game developers it can seem a relatively new concept. Software development, distinguished from game development, has more rapidly embraced the need for sustainability and standard sets of practices and approaches. This is due, in part, to software, generically considered, not being subject to many of the same restrictions and structures of secrecy that pervade the game industry. Even the menus presented to the "player"/designer of a game may be adjusted or have new options defined. This allows for new data, or experimental opportunities, to be pushed into the game. New possibilities can be created, default properties defined or adequately restrained for designers to then explore. Thus, even the most basic approach to providing designers the ability to explore a games system, can quickly become very complex. As more powerful and playful systems are added to the game development process, it is no wonder that these technological systems rapidly approach massive scales and massive complexity.

Prior to the development of the more integrated tools for *SM3* (detailed in World 4), VV's preproduction team relied heavily upon the debug menus of the game. "White room" levels were created, where artists and designers could add, or "spawn," newly created game characters to test animation sequences, behaviors, model appearances, and many other functions critical to the game. The majority of a game's behavior comes from locations external to its "code." Models need to be loaded, not based on hard-coded source, but based on design data from game designers. Artists need to be able to specify a range of textures applicable to a single model and the frames associated with animations. In many cases a single model may be used to define several different "enemies" within a game. What distinguishes each object may only be a shift in color values, textures, properties (e.g., attack, defense, vulnerabilities, immunities), or size. Each one of these can be experimentally defined through interactive design.

It is precisely the ability to play with the functions of a game to better understand each components relationship with the broader game systems that is so important. It is also the aspect of game development that is so difficult for small game companies to come to understand. In the case of studios based in India, working on outsourced projects, it was precisely these tools that were denied them, making the iterative, exploratory and playful process of game development untenable.

World 3-3: Rise of the Technical Artist and Tools Engineer

Box 3.2

#: SET DEMO_MODE 1
AUTHOR_DEV_DILEMMA: The vague connotation "communication" becomes a kind of stand in for a whole host of social, technical, methodological, and epistemological differences between members of a game development team. As such, it is readily identified in so many cases as the culprit that hounded a game development team.
Casey: Sure. How is communication between the different groups working on the project?
ENG_GRP_MGR_1: Communication is pretty good . . . Well, I wouldn't go that far. Communication is getting better. There have been times in the past when it has been poor. That is just one of

those areas that we need to work on. We have a group of folks on the west coast, and our communication with them is very poor. On a previous project we were working on, we had acquired the assets and software of some folks from this company when they went out of business. We had sort of a destructive and abusive relationship where each team would blame each other for things. It just got out of hand and was pretty horrible. So we are actually reaping the benefits of that now and trying to improve communication. It is a slow process. Communication is alright. We're taking new actions all the time. I mean, we have this guy TECH_ART_1 who is in the office with you. He's a technical artist. That is a position we didn't have until six months ago. There is now this person who can actually facilitate that communication between programmers and artists. That has been a really good thing. I'm hoping we'll do more of those kinds of things. We're trying different processes to figure out how to get teams communicating better.

AUTHOR_DEV_DILEMMA: New "disciplines" emerge precisely at those fault lines where the greatest friction occurs. The individuals that inhabit these spaces attempt to bridge or mediate across boundaries that make them inadvertent obligatory passage points. Yet, it isn't "communication" precisely that is being groped after. It is something more ethereal and difficult to put one's finger on.

Casey: Can you put your finger on the source of a point of pain? "This is where we have weak communication."

TECH_ART_1: I spend so much time communicating here. I spend more time communicating than doing anything. That is kind of the nature of where we are at in the project right now. We are in pre-, so we are making decisions about where things are going to go. It requires a lot of communication. I would like to see . . . I don't even know how . . . You spend so much time writing emails and posting on forums and having meetings, and everything. If there was just some way we could all mind-meld and get it done, in a more streamlined way. I don't know what that is though.

AUTHOR_DEV_DILEMMA: Ultimately, for developers, communication is about what makes good collaboration across disciplinary divides. Attempting to understand the underlying systems and structures that shape each other's understanding of a problem. Communication is about developing a clear commitment to understanding what another team member is attempting to do and how the realm of possibilities that make that possible or impossible can be bridged in ways that leverage the particular skills of each.

Casey: Are the distinctions and relationships between disciplines pretty clear?

TECH_ART_1: That has always been a clear distinction, and yes, my job as a technical artist is to straddle that line. But, that

(Continued)

is why I think I'm needed. There is a huge wall between artists and engineers. In many cases that may actually also be a physical wall, they are often sitting in different places. Most of the time it is a giant communication barrier though. So that is a big thing. Then you have management, who is also speaking a different language at the same time. Designers and writers speak a different language too. So, I definitely think there are very clear lines between the different disciplines.
Casey: Do you think that is a good thing?
TECH_ART_1: It can be frustrating when someone takes their disciplinary responsibility to mean it is theirs and theirs alone to discuss and decide. I think every area needs collaboration. A programmer is entitled to an opinion about how something looks and an artist is entitled to an idea about how something acts in the game, and everyone is entitled to thoughts about design and mechanics decisions. So think you need to have an openness for discussion, but I also think that roles and boundaries are good when it comes to decision making time and responsibilities. Everyone should be able to hear what others have to say, but I have no problem with DSN_LEAD_Spidey_1 being the lead designer and making the final decision on what gets put into the game. It would be absolute chaos if there was a team of people and we were "just going to do it."
Casey: What do you think is making it so difficult?
TECH_ART_1: I think unless you all sit in the exact same room, then you're going to have some sort of medium like that. But even if you don't, it is important for recording things. Email is really great for moments where you ask yourself, "Wait, what did that guy say?" So, even if everything happened verbally, but how do you keep track of what was decided? I think in some ways having everybody sitting in one open area has its perks, especially during preproduction. Once you get into production, you kind of want to just take your list of stuff and go into your hole and make stuff and not be bothered though. In preproduction though, you have to be talking to all of the people on the team.
AUTHOR_DEV_DILEMMA: Collaboration and communication are key elements throughout the creative process of game development. Understanding the structures that lie behind engineers' often oversimplified "no" can be frustrating for artists. The desire for a particular artistic effect or capability is also framed by structures that are not understood by engineers. The ability to see those conversations in a way that enables the identification of common frames or an understanding that bridges them is precisely what has made technical artists and tools engineers such a critical component of modern game development.
#: SET DEMO_MODE 0

Game development has seen a dramatic shift in the last eight years. The amount of digital storage space available for developers to use has risen dramatically and the subsequent expectations of players and publishers have risen as well. This has meant that in many cases, significantly more content can and must be placed into a game. More levels, more models, more textures, more sounds must be created to fill the available space on shipped disks. This has required a significant shift in how developers approach game development and the pipeline has become much more important since it is the path through which all those elements must flow to become part of the game. The process by which a particular game asset (sound, image, model, level) is placed into a game and how it may be interacted with by designers and engineers further down this pipeline has only increased in importance. Specifically, the turnaround time for an artist or designer to see something in the game such that they can ensure that what they've constructed in *3D Studio Max* or *Maya* indeed looks as it should inside the game is no longer a luxury, but a necessity. In the case of designers, this means that a player moves as anticipated, a level is paced in a way that is playable, or that "out of bound" areas cannot be accessed.

The pipeline has become much more complex over this time period, and its two main laborers have become the "technical artist" and the "tools engineer." Each serves different purposes, though in smaller game studios the technical artist and tools engineer are often the same person. The technical artist is an accident of history, with many technical artists having moved from computer science programs during college careers to artistic degrees when the myth of the programmer-only game development haze has cleared from their eyes. At the same time, these individuals had in many cases acquired an appreciation and understanding of how software engineers understand the world around them. This kind of overlapping knowledge made them uniquely prepared to interface between the perspectives of artists and engineers. Through a similar exposure to programming and art, some technical artists without formal programming backgrounds simply emerge in small game companies by necessity, making scripts, toolbars, and other utilities that speed the process of getting their work into the engine so they can see it. They are the saviors of other artists when things don't go quite as anticipated because they have the knowledge ands skills to either fix a problem, or push back on the engineering team to explain why or how something is broken in a way that other artists may not.

The technical artist serves a crucial role as a kind of mediator between engineers and artists. In videogame development, engineers and artists largely speak a different language and focus on different aspects of the game development process; in short, they come from different epistemic communities (Knorr-Cetina 1999). Technical artists will often understand the idiosyncratic differences in how artists and engineers talk about the underlying systems of a game and has thus emerged as an interface between the disciplines of engineer and artist. What makes technical artists particularly interesting is their emergence specifically at this disciplinary fault line. Artists and engineers have different understandings of what "counts as a good question, an interesting mode of inquiry, way of teaching and learning, and the infrastructure needed for pursuing these emerging forms of knowledge" (Traweek 2000, 23).

Anthropologists of science and technology have demonstrated that it is frequently at these fault lines of disciplines where the most interesting and critically important outcomes occur (Traweek 2000). Historians of science have similar findings, noting that it is at these sites or "trading zones" where "creole languages" emerge and local coordination and cooperation can be worked out in practice (Galison 1997). Intimately important to this process, however, is the ability to get at the underlying systems that historically situate our object of concern. This access is significantly limited in the videogame industry. For the collaborative process to really function, it is important for "open system analysis" to be possible, a process dependent upon the historicization of the object of concern (Fortun 2006). Because of the emphasis placed on closed systems and closed collaboration, it becomes difficult to historicize: in this case, the very act of creative collaborative practice in the game industry.

The technical artist is a nomad, fully at home in neither one community nor the other. Technical artists, as mediators, often find themselves attempting to bridge the incommensurable. Emerging at a fault line means that they often find themselves at points of extreme friction, which are precisely the points at which technical artists contribute to the pipeline (which serves as its own kind of fault line mediator). As technical artists put in place scripts, exporters, optimizers, toolbars, documentation, and other elements that facilitate smooth transitions from artists' tools to the game engine, they grease the wheels of videogame production by replacing communication and understanding between artists and engineers with technological systems. These systems make invisible the significant labor required to bridge disciplinary understanding of videogame development. These are also the pressure points that break during periods of intense stress and instrumental activity of artists during game production.

Placing technological systems at fault lines makes them manageable in a different way. "Good, usable systems disappear almost by definition" when they facilitate communication gaps, or minimize them (Bowker and Star 1999, 33). For game developers, this means that pipelines disappear into the background. Perhaps more important, good usable systems mediate effectively between different disciplinary perspectives. When those systems fail, they can have "bug" reports filed on their failure. It creates a kind of abstracted separation between the human and the machine. Staging experts and systems at the junction more effectively manages the gap.

During the preproduction phase of SM3, the lead technical artist came to me looking for advice. He had been advised that "communication was not so good" between the art group and engineering group. Since I had been observing at the site of turmoil, he turned to me asking for guidance. We talked about the situation for most of a morning, which is a massive amount of time for a game developer at work. We collaborated, discussing both of our observations, and came to decide that the real fault line was based on disciplinary difference of understanding what makes the project tick. We identified four different ways that the project was being viewed, illustrated by Google Maps and Google Earth. Figures 3.1 and 3.2 represent different perspectives of the same problem and its subcomponents. They were separated by scale (level of detail) and in content (art or code). We found that, as one might expect, artists were typically interested in understanding the game in a way that favored artistic aspects (represented by the

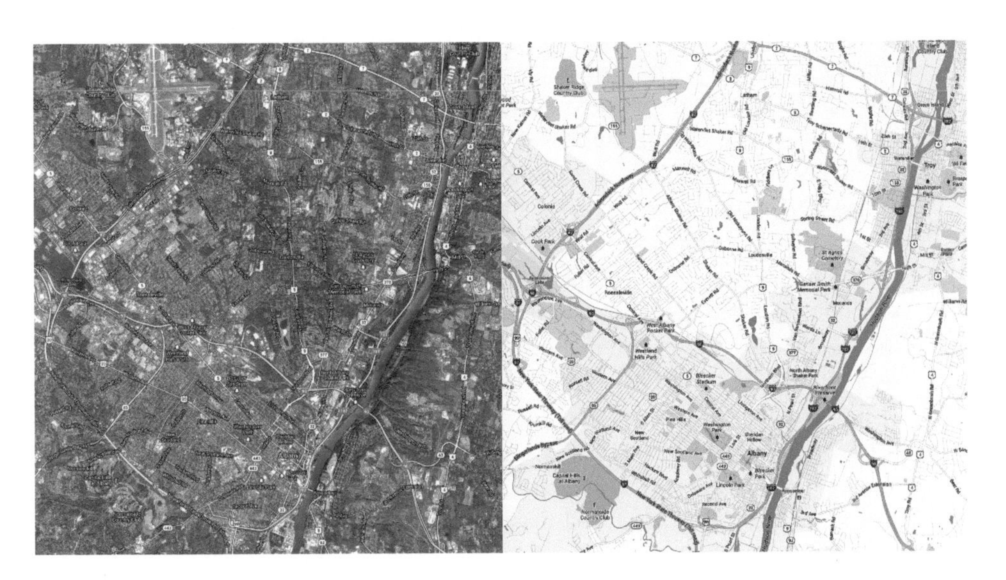

Figure 3.1
High-scale images of art (left) and code (right) conceptions

Figure 3.2
Low-scale images of art (left) and code (right) conceptions

satellite images). Engineers were primarily concerned with implementation (represented by the road maps).

One possible resolution would be to simply lay the maps on top of one another, making a hybrid, a popular solution among many developers (aka "the mind meld"). Unfortunately, this would not be a helpful solution. Attempting to teach your artists about all of the engineering aspects and vice versa would be both cumbersome and likely impossible. The utility of specialization is that they should not need to know everything the other knows. Different scales and content is useful. Homogeneity was not the goal; communication was.

The higher scale "engineering" or "art" map (figure 3.1) illustrates the viewpoint of the engineering or art lead. Based on the map, it is obvious that this person's greatest knowledge will be the overall functionality of a system. They will likely have less knowledge of the system's lowest level of functionality (represented by figure 3.2). Nor will a lead artist have the details of lower scales. But these acknowledged differences are necessary for the project to come to completion.

The solution, as the lead technical artist saw it, was to encourage all parties to understand the utility and "correctness" of each interpretation, or to understand the need filled by that perspective (which was often lost in the other perspectives). While engineers might control the flow of art assets into the game, artists wanted some information about why an engineer was saying "no." Engineers also needed to understand why artists were attempting to create certain effects or models. The lead technical artist therefore encouraged both groups to understand the differences in their viewpoints and scales people were working at, providing developers with a new language for discussing collaborations helped them work together.

Of course, the question remains whether this approach changes anything: does providing new cognitive tools matter when it comes to the production schedule of a game? While the hopeful answer is that over time the process will change and become more collaborative, the short-term answer has been that it has not yet altered how people work.[2] In one case, the producer of a project was meeting with the executive producers from the publishing company for their upcoming game. The publisher commented that the buildings in the Bronx, despite being to scale, did not look tall enough. The producer, wishing to please the publisher, went to an artist and asked that he scale all of the buildings in that block so that they would be taller, a task that was relatively easy for the artist with the tools that had been developed. Unfortunately, since the producer had bypassed the art lead, no one thought to check that all of the collision data (the data used to determine which objects in the game a game character "collides" into and which impede their motion, such as the ground, a rooftop, or a wall[3]) would also need to be altered. Rather than impressing the publisher, the resulting demonstration, in which characters passed through walls and fell through the ground, caused the publisher to doubt the capabilities of the development team.

In some cases, studios have attempted to automate the technical artist's mediator role. Engineers on a development team occasionally attempt to create engines capable of bringing in nearly every model and texture format, in the hopes that designers and engineers can then interface with the creations of artists without the constraints of waiting for and taking time from another developer. The difficulty is that such an approach does not recognize the kind of complexity and nuance of technical artists' skills. Technical artists often recognize problems in advance and encourage artists to follow sets of practices that further down the pipeline often result in fewer difficulties, simply based on their overlapping understanding of art and engineering. For example, technical artists will often encourage animators to use nearly exclusively the more complex "quaternion" system for dealing with rotation, knowing that the underlying engine of the game is less likely to encounter errors with such a rotation than with the simpler "angular" rotations.

Thus, the experimental systems created by technical artists along the fault lines between artists and engineers are not simply labyrinthine (as we previously defined it to mean constructed on the fly). Rather, technical artist systems are rooted in art, craft, and science. This further complicates our understanding of technological working, "as craftpeople informed by abstract knowledge" (Barley and Orr 1997a, 12). Work in this context is art, craft, science, and interest in collaborative practice. Collaboration requires interest and overlap in realms of understanding, furthering the idea that for game development practice to function, more people, not fewer, will be necessary to keep the "black box" functioning.

The tools engineer, much like the technical artist, has been an accident of history, rather than a deliberate shift of the industry. Tools engineers were typically engineers who found themselves watching artists, designers, or engineers continually making the same "mistakes" over and over getting things into the game. Tools engineers' core role seems to be helping others manage the chaos of game development.[4] This has led to their construction of custom generating tools, many of which may have been previously specified with the editing of TEXT, INI, or XML files. Custom editors for games are the creations of tools engineers. Perhaps unfortunately, tools engineers have also become the masters of build systems and must frequently perform numerous tasks and integrate the persnickety compilers and tools developed by console manufacturers with little apparent regard to usability.[5]

Ultimately, however, each one of the tools engineers and technical artists—has made it a goal to create game development systems that respond rapidly to the work of the developer. Adjusting a slider and being able to see the change in particle system behavior is much more intuitive than the pre-tools-engineer days of editing external files or code directly. Dropping a new texture onto a model or selecting it from a drop down menu is far more responsive than having to consult with an engineer to modify the code loading a texture and wait for a new build of the game. Clicking a single button to perform a model check, export, and load into the game engine takes less time than following a checklist. These hybrid developers and their resulting systems are actually extensions of the World 3-2 discussions of debug menus and consoles within games. The interactive systems' objective is to provide flexibility and make the lives of developers easier. As with debug menus and consoles, the proprietary tools and pipelines developed by technical artists and tools engineers are rarely available for gamers and are built and available for preproduction and production phases of development only.

While technical artists' positions can be best thought about as occurring at fault lines between artists and engineers, where friction is managed through the development of practices and systems, tools engineering is somewhat more systematic and done with instrumental work/play in mind. Tools engineers function at the "trading zone," which can be thought of as a "domain in which procedures [and technologies] could be coordinated locally even when broader meanings classed" (Galison 1997, 46). Put another way, the systems developed by tools engineers serve as the kind of "creole languages" developed by those living and working in trading zones.

> What is crucial is that in the logical context of the trading zone, despite the differences in classification, significance, and standards of demonstration, the two groups collaborate. They can come to a consensus about the procedure of exchange, about the mechanisms to determine when goods are "equal" to one another. They can even both understand that the continuation of exchange is a prerequisite to the survival of the larger culture of which they are part. (Galison 1997, 803)

In this negotiated exchange, then, tools are very much a monetary commodity within videogame development companies. For the management of a videogame company, its tool chain is intellectual property that provides a significant edge in the act of videogame production. For game developers on the ground in those organizations, these technologies have "intention, purpose, and priorities," (Galison 1997, 804) that enable and constrain certain activities over others. Tools, because of their negotiated character, could also be conceptualized as a kind of game development "boundary object," (Leigh Star and Griesemer 1989) though I think the metaphor of the "creole" language is much more accurate, given the often heated conversations that take place in videogame companies.

The tools created by tools engineers in many ways also do the work of managing friction between different disciplinary understandings of videogame design and development. At the same time, tools are much deeper systems than those created by the technical artist. While the work of the technical artist often disappears, the work of the tools engineer is often very visible within the organization. The aspects of a pipeline created by a tools engineering team may very well be productized and sold to other game development studios. At that point, however, the pipeline tools are then often divorced from their broader context and from systems developed by technical artists that further smooth and enable the system to fit well into the game development process. In other words, the tools engineers create invaluable resources that are much less useful out of context, just as a creole is less useful (and often incomprehensible) beyond its trading zone.

Separating such tools from the game they were developed to create often requires significant work on the part of engineering teams, artists, and designers. Any game development studio that has used a commercial game engine will attest that it is not a trivial task to integrate the engine into the company's workflow and make it work for the game design as initially conceived. After all, videogame development tools and systems are highly experimental. They are created with specific games in mind and they therefore significantly shape the kinds of games that can be created with them. The tools are also continually being instrumentally worked/played by technical artists and others who smooth the process of fitting these tools into their broader social/technical/artistic context.

World 3-4: Fault Lines, Fault Lines, Fault Lines Everywhere

As an engineer, artist, or designer proves their abilities and gains experience, they will more than likely begin to move into either a "lead," "manager," or "producer" role in a game company. This is assuming that the studio has grown large enough to warrant these roles. For the most part these distinctions were quite similar between US- and India-based studios. In smaller studios, every employee fulfills part of these roles out of necessity. While the basics are similar, leads, managers, and producers[6] are different and distinguishing among them is important.

A lead is frequently an artist, engineer, or designer who has proven their ability to produce quality work and exhibits some leadership characteristics. Leads remain responsible for producing design data, code or art assets in addition to their leadership role. Yet, leads must also maintain a higher-level understanding of what the entire team is working on and how it relates to the project more broadly. They typically come up through the ranks of an organization. Their responsibility is to represent their group's interests in team meetings and planning sessions. They need to adequately report back to producers and management on the status of a project and therefore tend to work closely with their teams to ensure adequate information flow. Some leads seem to have extended the role of lead artist, engineer, or designer into that of a project lead. In some cases this is effective. In other cases, management who come to their role via a group management role will often find themselves micromanaging certain tasks of the development process, unable to let go of tasks that they likely had been responsible for in their previous role as lead. Managing a project or group of a project, though, separates the lead from managers, for leads tend to be responsible for a single project.

Managers, on the other hand, lead multiple teams through multiple projects. Also unlike leads, managers generally do not take part in the production of games. They may be involved in the production of internal resources for their teams, but their overall goal begins to shift to ensuring the long-term success of their "groups." Most managers, like leads, have risen through the ranks of the organization and are frequently artists, engineers, or designers who have been with the company for a significant amount of time and have been willing to move into managerial positions. Some developers choose to remain leads rather than moving into management positions. Ultimately this means that very few members of a game studios "management team" have any management training. Some organizations will work to train in key areas, such as communication and leadership.

The role of the producer is the management of a project or, for an executive-level producer, several projects. In the end, they are responsible for the overall quality, profitability, schedule, and effective production practices of a given project. Producers must understand the scheduling and staffing needs of a project and ensure that milestones are met. Much like leads and group managers, producers also suffer and benefit from having been active game developers. It gives them an intimate understanding of how games are created and developed, but it frequently results in lack of management training and sometimes with an actual disconnect from developers because their attention may be focused more particularly on those areas with they are most familiar. The producer is the person who ultimately, at least organizationally, is considered responsible for the relative success or failure of a game. Producer roles generally seem similar from the United States and India, though the number of producers in India was diminutive, as there were fewer "end to end" or complete game development projects underway.

The leads and the producers often occupy fault line positions much like technical artists and tools engineers do. The transition for a lead or producer from one realm to another positions them as mediators between different disciplinary perspectives. While each may tend to favor their previous point of view, moving into roles that demand a different kind of understanding forever corrupts them. What is interesting about this is that it results in a kind of misunderstanding of where problems come from. They become, as we've seen before, generically labeled "communication" problems.

In creative collaborative work, communication is actually more about understanding perspectives other than one's own. Communication in this context is attempting to understand, and perhaps more importantly, caring about the perspective of those other members of a team with epistemic frames different from one's own. This epistemic bridging and communication takes time and is one of the more prevalent activities of game developers. Communication for game developers is about care and work, which occurs in the much dreaded, meeting.

In the written descriptions of the roles of engineers, artists, and designers, the phrase "when not in meetings" was used as a recursive feedback placeholder. The interactive organization loops in on itself. Management schedules meetings to better understand what is transpiring at each level. Leads will have team meetings, producers will have meetings with leads, and managers will have meetings with their disciplinary groups. Studio heads will have meetings with managers and producers. Entire teams will participate in company-wide meetings. Information flows in both directions, but primarily it flows up the chain of command. Management time is dominated by meetings and ensuring that things are functioning properly—scheduling meetings and also participating in them. The importance of face-to-face meetings, for which US or Indian developers make time outside of development hours (they will arrive early at work or stay late for meetings), "manifests a delicious contradiction; work becomes more dependent upon workers' abilities to create close social relations at the same time as globalization inhibits their construction" (Hakken 2000a, 771).[7]

Some leads, managers and producers borrow from more formal project management techniques, yet most are assembled in an ad-hoc fashion. Even those that make strides to bring new tools and approaches into the workplace will, given time and pressure, often revert to micromanaging or step out of management roles out of inexperience. This is exacerbated as corporations globalize and become a seemingly "complex tangle of remotely related parts. . . . both tightly coupled and dispersed," (Fortun 2001, 93), a recipe discussed in this level's boss fight as particularly problematic.

The process of relating the parts to the whole is often problematic in management and yet is something that videogame developers constantly expect of their players. Integration and interrelation ought to be something of a specialty for this community. The numerous data sources, such as VCSs and logs from the build systems could be leveraged in ways that provide leads, managers, and producers with new tools to make sense of the work happening all around them. The game development process cannot just remain a complex tangle, given the kinds of feedback loops that create dependencies and relations that ultimately result in a kind of feedback fetish, where feedback and response themselves become goals, rather than a means of understanding how the system functions. In part, the mess between and among management realms is a failure to see understanding the process of game development work in and of itself as worthy of attention within game companies. Leads, managers, and producers are left to their own devices in developing tools, metrics, and practices by which to understand how a team is working.

World 3 Boss Fight: The Fun Part Is Over Now

It is strange that game developers can constantly be focused on creating meaningful experiences for players, which often requires the presentation of extensive amounts of data in abstract ways—data related through numerous feedback loops that allow players to make decisions. A renewed focus on tools and the pipeline and providing flexibility for development teams to understand how their own systems interact has been only relatively recent. The basic truth is that videogame developers need the stability and access that more open systems bring. The details, however, attest that this requires a fundamental shift in developers thinking about what needs to be kept secret and those things that can be more widely shared such that they can be leveraged to improve the working practices that mobilize their daily lives.

Within the game industry there are emerging standards and half standards and rules of thumb, but there has been very little work toward real industry-wide standards or best practices. There needs to be some base level that is approachable and deployable broadly. This is where efforts, such as Microsoft's XNA fall short. The continued lock-in to largely proprietary systems like C# and DirectX contradict the call for standardization and open access. There are numerous "industry standards" that end up competing with other proprietary standards. Developers frequently pick and choose based upon those with the newest features rather than on those that have open standards. Having standards does not mean that individual companies cannot go above and beyond those industry standards. Nor is this to suggest that only Nintendo, Sony, Microsoft, or some other large, well established company must be the only player in town. Analysts have shut down calls for standards in the past through misinterpretation or poor "analysis." Periodically a call for some kind of standardization goes out in the game industry (Waters 2007). The response is typical, "I do not think the industry will ever resort to one console. It would be bad for the industry. I could understand the argument for a single development standard, but not a single hardware standard. . . . However, in regards to a single console, it would hinder innovation and consumer choice" (Wen 2007). Standards are instantly conflated with a single console, which is not the point.

Either that or the idea is instantly tossed out as impossible, as a developer did when asked about standardization: "The fact is, as long as Sony and Nintendo are alive and kicking, one platform will never happen. If there were one platform, the manufacturer would have all of the leverage, unless it offered open architecture. Not likely. Nintendo and Sony [would] both insist upon a proprietary standard. Microsoft has a proprietary online business. It sounds wonderful, but so does world peace" (Wen 2007). A culture of secrecy internalized by the developers themselves means that, not only do they not push for standards, but we shout down calls for such openness. Developers who would benefit from fewer secrets condemn any call for better work practices.

Ultimately this level's recommendation is dependent upon implementation. Standards that span platforms could be developed, but developers would have to be willing to avoid platforms that shirk those standards. A more modest intervention would be for developers to consider fewer aspects of their work worthy of secrecy. Developers need to demand that process and source level knowledge of development practices need not be included under NDAs, allowing developers to contribute back to OSS initiatives that they draw upon. Developers have to be willing to take a stand and support those more open technologies, which ultimately will improve their ability to do game development work. Hegemony, the organizing principles by which society agrees to the priorities of the powerful works by seeming like the natural order of things. But it's not the natural order of things, nor is it simply about the status quo or domination. Hegemony embodies a kind of dialectic between coercion and consent. And it is subject to resistance and recalcitrance. Developers need not simply roll over for those corporations that make dictates about how game development needs to occur. Experienced developers, in particular, can push back against restrictive practices that make game development work dysfunctional. And they can ask for change. Counter-hegemonic projects can force accommodation to their demands. Organizations in positions of power cannot simply say "no," for in all likelihood an open revolt could occur among developers. In fact, this has occurred as developers moved away from developing games for the Super NES and N64 to those less restrictive agreements surrounding the Sony PlayStation. Developers have the opportunity to push back against NDAs and licensing agreements in ways that could enable greater stability. They can help create a more open, less secretive industry in which to creatively collaborate, work, and play.

New tools, practices, and processes can be developed from a more open foundation. This is precisely where Free/Libre and Open Source Software (F/

LOSS) has proven its ability to support unified technologies across numerous hardware platforms and the production of resources that developers can draw on to reduce their dependence on self-developed code. F/LOSS has dramatically influenced much of the broader software development community, but because the game industry has assumed itself different, it has not yet been able to grasp the importance of openness for pushing technologies, practices, and stability further. The constant re-creation of software often results in bugs or deficiencies due to the process of developing software. F/LOSS addresses scalability (the ability for similar APIs to work across numerous devices ranging from full-scale computers to very limited custom hardware) in ways that the videogame industry could dramatically benefit from.

Ultimately, much of what game development companies pay for now as "middleware" is software that could be more effectively developed in an open and cooperative manner. This would also allow developers to work with similar tools even in locations where licensing agreements are unattainable because of cost or due to restrictions by hardware manufacturers. And this approach would allow developers the ability to retain more of their work when licensing was provided.

This is really an appeal to open up the production networks of the game industry wide enough to enable collaborations that extend beyond the walls of single studios, publishers or manufacturers. It is an earnest call to enable work to flow more broadly into the worlds of free and open source software in ways that are not simply extractive, but give back to those communities in new ways. This call to change is not about asking the game industry to become completely open or porous, but capable of providing interfaces that may very well make the daily lives of developers better. F/LOSS developers are just as excited at seeing game developers draw on and contribute back to their projects as any other community.

What is at stake here is not the "giving away" of free DevKits, but rather the opening up of the SDKs that allow developers to create games for these systems. Production pipelines for designers and artists will emerge around these SDKs. To make this possible, the enormously secretive console manufacturers will have to provide some mechanism by which games can be tested without DevKits (these devices are already being provided outside the law, and will be used in the implementation of the game described in World 8). This would not spell the end of multiple systems; rather it would assure their place in the market, for, surely aspiring developers would be more likely to purchase "approved" Nintendo, Sony, or Microsoft versions of each device.

Many independent developers or F/LOSS developers want to be included in the networks of the game industry and have simply been told, "no," or often simply nothing at all. In stranger cases, F/LOSS developers have often, unknowingly, had their projects incorporated into commercial games with no reciprocal relationship being formed. There is clear value for game developers in leveraging these open technologies, but little way for them to contribute back to that broader "commons." As can be seen from the rapid rise of Apple's iOS for games, it is clear that there has been a built up need for change in this space. It should be possible for game console manufacturers to maintain control while also increasing the openness of their systems. The more interactive DevKits could be made available only when a developer has gotten approval to move forward with a manufacturer or publisher. That approval should not prevent a developer from pursuing development on a platform, however.

Box 3.3

#: SET DEMO_MODE 1
AUTHOR_DEV_DILEMMA: Cultivating experienced developers requires that aspiring developers be able to gain experience in actual production processes. As currently configured, that is nearly impossible given the kind of control exerted over the hardware that dominates the game industry. Developers are asked instead to work in hypothetical situations that are often completely uninformed by the reality of game development.
Casey: You were talking about some of the production difficulties on *SM3* . . .
DESIGN_LEAD_1: Yeah . . . Generally our production methods on *SM3* sucked. Principally just test tracking and things like that. We went through several versions and setting on something that relied on inexperienced production coordinators to track it all, so that took forever. The team was just inexperienced overall. It was a very green team. We had a lot of co-ops, especially on mission design and they had to work a lot harder to make up for their inexperience. But that was a staffing problem, which went back to a staffing freeze several years ago, which had a loophole for non-salary folks, basically co-ops. Some of them were awesome, and we hired them. Others went to work at places like EA. Even some of our leads were inexperienced.
Casey: So where should that experience come from?
DESIGN_LEAD_1: Well, my opinion is that the [Nintendo] DS is the perfect proving ground for growing people. We need to be growing people through DS and then having them come on the console side with that experience.

```
AUTHOR_DEV_DILEMMA: There is a strange mantra that dominates the
game industry. If you want to make games, you need to go make
games, which makes sense. If you want to write, you should go
write. Yet, the very process by which one makes a game is made so
opaque, and even more so if thinking in terms of modern games. It
isn't about technological standards so much as it is about stan-
dards of practice.
#: SET DEMO_MODE 0
```

The practice of videogame development needs standards. The continued opacity and closed character of the industry has lead developers to continually forge the same path over and over again. Having once traversed a course, developers have been unable to share in detail these routes because of restrictive NDAs and licensing agreements. Console manufacturers and publishers continue to hold tightly to the technological and legal keys to the means of distribution, and will continue to do so unless the industry itself demands change. The obsession with control over the production of games is a relic of the past that hinders developers and the industry. For the industry to truly mature this tight-lipped control of information must change. Further, such changes and the resulting maturation of the videogame industry will lead to improvements in quality of life (QOL) and sustainability, the major concerns for my US- and India-based informants.

Perhaps more important, game studios need to learn how to and be allowed to "carry string with them" during the construction of their labyrinths. This process has no technological fix. It is a deep-seated social component that necessitates taking time to reflect on the activities of game development, and not simply at the end of a game title's development. Slowing down long enough to talk about how and why something should be done gives development teams a better understanding of why they are progressing down one path rather than another. More connections, detailed sharing, and openness will result in more mature videogame studios being able to work within the numerous structures of access. Ultimately however, this kind of revised videogame development work requires significant change in the numerous structures that shape the lived realities of videogame developers. These changes developers must demand themselves to make new kinds of videogame titles possible. For too long developers have allowed the broad forces of the videogame industry to shape their communities. These are castles that must be hacked and MODed.

Production: Let's Go Make Stuff!

World 4: Interactive Game Development Tools

Box 4.1

#: SET DEMO_MODE 1
AUTHOR_DEV_DILEMMA: As we saw in World 3, game development relies on numerous tools (software systems) that are often ancillary to the actual game being developed by a team. These tools are by themselves complex software systems that must be created and maintained by teams of tools engineers and technical artists. This further complicates the already muddy understanding that many people have of game development practice.
Casey: Do you like sitting in between the artists, designers, and other engineers?
TOOLS_ENG_2: I do. That is exactly where I want to be.
Casey: What do you think are the main things that people don't understand about what you do?
TOOLS_ENG_2: It is really hard for me to explain my job to just about anyone else. Even just saying that you're a programmer, not in the game industry, people will respond with, "Well, what do you do?" But people don't take it seriously because it's videogames. Or, people will think that all you do is screw around all day. I try to explain it with concrete examples, like trying to explain what a level editor is or what a level designer would do with tools I've created. That is the easiest way to explain what a tool might be. But people don't understand anything other than the final end product.
Casey: The box on the shelf?
TOOLS_ENG_2: Yep. You can explain what an artist does more readily. To a certain extent you can even explain a regular engine programmer, but a tools programmer . . . They just don't . . .
AUTHOR_DEV_DILEMMA: This ambiguity between what is the game and what is the software that surrounds and supports it makes it difficult for many developers to explain what it is they do and for many, this can make it seem like their labor associated with a games development disappears behind those roles more directly identifiable as having participated in a games creation.
#: SET DEMO_MODE 0

World 4-1: Engineering Interactivity

Once a game has reached production, most of the tools and processes have been nailed down. The everyday operations are supposed to be less experimental and more about actually creating all of the elements planned for a game. Of course, if preproduction was unsuccessful, then this becomes more complex. Interactivity is an important aspect of how game developers understand the relationship between their work and the videogames. It is similar to the "meaningful play" game players discover as a result of their actions in a virtual space (Salen and Zimmerman 2004, 33–36).

Throughout the production process of *SM3*, game development pivoted on interactivity as additional tools and systems were developed to help each group working on the project understand the relationship between their work and the work of others. It was also during production that the number of people working on the project dramatically increased. Additional artists, engineers, and designers were added to the team to create the mountain of art assets, code, and data that would become *SM3*. In this world, we examine these interactive intersections between engineers, artists, and designers. The world concludes with an examination of the complex issue of "the build" where one demanding goal is keeping a game's voluminous system synchronized across the activities of numerous developers. Finally, because game development is a creative collaborative practice that bears little resemblance to how it might otherwise be imagined or portrayed in popular culture I return to the imagined realm of what game development is thought to be to erode myths that hinder industry growth and maturation.

Throughout production, the engineering team of *SM3* was responsible for implementing the various systems that they had laid out and prototyped during the preproduction process. Typically, during this time of the production process, engineers can frequently be found in front of their computers, in front of a white board with other game developers, talking in person with other developers, or reviewing code written by other engineers. Time spent in front of the computer screen can be spent "working" or writing game code or documentation for game code, or it can be spent managing massive numbers of emails from across the company, personal emails, instant messages from co-workers, some work related, some not, and instant messages from family and friends. Occurring simultaneously are web browsing looking for documentation, reading up on industry news, or pursuing personal interests. Meeting reminders pop up on the screen courtesy of Microsoft Outlook or Mozilla Thunderbird.

Most of this time, however, engineers are in front of their integrated development environment (IDE), the tool designed for creating software. If they are writing source code, then typically their code will dominate one computer screen, and documentation will fill the other. Engineers pour over documentation determining how to take abstract concepts or solutions to problems and make them function in whatever environment or on whatever system they are developing for. While similarities will frequently exist between systems, oftentimes specific code must be written or rewritten if being developed for different systems. Just as code written for Windows is different from MacOS (both are operating systems or "platforms"), code for ODE is different from code for Havok (both are physics APIs) and DirectX is different from OpenGL (both are graphics APIs), despite the fact that similar conventions and goals frequently drive each system.

At the core of engineering for games is the idea of managing a game's state. This includes both the running simulation of the game's rules as defined by designers, and the maintenance and putting into motion of the art assets created by artists. Because this process is frequently unknown, or at least experimentally defined, engineers spend a great deal of time using the debugger (a component of their IDE); observing what a piece of code is doing; attempting to figure out why a piece of code is not functioning as they intended or believe it ought to function; or correcting mistakes made in the process of translating their own ideas into code. There are two productive ways of thinking about this process of determining the functionality of a system through experimentation. The first is the "mangle of practice," where there is a "temporal structuring of practice as a dialectic of resistance and accommodation" (Pickering 1995, xi). The second way is about getting better at understanding resistances. "Listening to noise and transforming it into a signal depends as much on acquired intuition" (Rheinberger 1997, 134). The experience of a developer "organize[s] the experimental gropings" (Rheinberger 1997, 134) into a functional piece of code.

The experimental process of debugging can be time-consuming and tedious, yet it exemplifies how interactive systems begin to inundate the game development process. One conversation among my informants highlights the ways in which interactivity enables production:

I look up from my computer screen as Eric begins yelling "fuck you" at his computer monitor. He's in the debugger, looking at some of the data moving through the game's engine. There's nothing pretty about this. It is numbers, strings, and source code. From across the desk David looks up and has this conversation with Eric:

David: Did that function fix it?
Eric: No.
David: Did that function help?
Eric: No.
David: I assume you figured it out?
Eric: No.
David: Do you want to talk about it?

At which point a math discussion ensues at the marker board. (Informants and O'Donnell 2005c)

This conversation is an excellent example of how the interactive tool, the debugger, in concert with a fellow coworker enabled a conversation about what was going wrong at the core of a game in development. The IDE provided the tool that made visible elements of the underlying systems of a game, which were clearly not working as expected, and allowed a conversation to ensue that later solved the problem. IDEs allow engineers to quickly and easily make changes to their code, recompile the game, and either see changes in the debugger or observe the behavior of the game.

It is this kind of experimental feeling out of functionality and behavior that has led some to refer to "successful practice" as "depend[ent] on trial and error or on local and contextual knowledge, then that too has generally been acceptable to most engineers" (Whalley and Barley 1997, 30). Both interesting and problematic, the "view of skill as having an improvisational quality is in stark contrast to lab managers' and administrators'conceptions of technician skill" (Scarselletta 1997, 207). The approaches of engineers have been thrust on numerous other actors who may not work well within these systems or approaches. Again, this seems to reference the "gate-keeping" aspects of engineering work in game development.

At the same time, the improvisational feel can be the first step down our path of fetishizing interactivity. Interactivity is useful, it provides feedback to the developer, yet it can also replace the process of attempting to understand the underlying system. Especially as timelines become tighter, which is commonplace during production, the desire for a system that perfectly responds to every tweak made by a developer becomes tempting. In some cases changes are made specifically as an engineer experimentally progresses toward a solution to a problem. In other cases changes will be made with little analytical foundation (for example, the addition of a "-" sign before a number or the adjustment of a constant number in the source code, all toward the end of a desired outcome). This practice is much more common among younger developers, many of whom have not had the experience of such actions back-firing on them—too many minute changes ends up in an unsalvageable or unreturnable past and represent death by one thousand cuts. Regardless of how the changes are made—as part of a methodical experimentation or of massaged code—this constantly changing set of what represents a game's underlying code means most engineers have adopted some sort of version control system (VCS).

The VCS can be as simple and inexpensive as a "shared folder" on the network that contains the latest working version of files used to build a game, or as complex as proprietary systems (such as Perforce, produced by a company with the same name). One key aspect of a VCS is that it tracks changes over time. While a shared folder cannot do this on its own, it can be approximated by compressing and saving the shared folder periodically along with a date. Other systems such as the open source CVS or SVN (Concurrent Versions System and Subversion, respectively) are free, though their integration with tools such as Visual Studio and Max are minimal.[1] An engineer using the VCS will "check out" a file of the version control system when they are working on it. This is visible to other developers who may choose to not work on the file at the same time, though in some cases the system may prevent others from working on it at all. When a file is "checked in" after work has completed, an engineer will typically make note of what was changed in the file.

"The build" is closely tied to the VCS. More than any other technological system in game development, it is what everyone on a development team seems constantly aware of—so much so that it becomes a kind of obsession. To visually inform anyone waiting on the build, traffic lights or strands of Christmas lights are connected to the build machine to provide instant visual feedback: green is good, yellow not so good, and red means "broken." Sometimes the breaking of the build is accompanied by an audio alert, the sound of an explosion or the screeching of tires. A Windows taskbar pop-up generated from figure 4.1 delivers warnings or errors to the user. The grey area surrounding the engineer's visage, originally magenta, is used to make those areas transparent.

Ironically, the kinds of errors displayed are machine-generated, rather than the kinds of words the engineering team lead would actually utter. For example, if a contributed art asset cannot be processed by the art pipeline tools, a message indicating the file's location and the associated error would be displayed in the bubble. When combined, the build and the VCS allow for automated builds and testing systems to keep track of the relative health

Figure 4.1
The image used to deliver warnings for users of the build system

of a game's progress. These automated builds also dramatically cut down on the amount of time necessary to "build" the game.

Figure 4.2 was generated from the massive number of files that it took to create the AAA ("triple A") videogame title, *SM3* for the PS2, Wii, and PSP. Because of the large number of files, the build can take a very long time. Automated build systems simplify this by having the latest version of processed files pregenerated, allowing an engineer, artist, or designer to only generate a small number of files to see the latest version of the game along with their most recent changes. Once changes have been tested, they are checked into the VCS, at which point the build system will incorporate them into the next build cycle—at least in an ideal world.

Engineers and, specifically tools engineers, are responsible for developing the software that combines the efforts of artists, designers, and everyone else into a playable game. These software systems read in the art assets produced by artists and the scripts, levels, or other pieces of data produced by designers. Even input read from controllers, keyboards, or mice is passed through the data produced by designers. The knowledge of when to play a particular sound file or display a particular model must come from somewhere, and frequently this is not hard-coded into the game code; it, too, must come from design data. Parsing data or files requires a detailed understanding of their format. Sometimes they are as simple as text files that can easily be parsed by reading the characters from the file and interpreting the information. In other cases the format is more complex. Binary, image, or sound data are, for all intents and purposes, unreadable by humans, though we may recognize these kinds of data when we see them. These file formats are often more multiplex, sometimes containing compressed data that must be uncompressed prior to its usage in game. This is the process

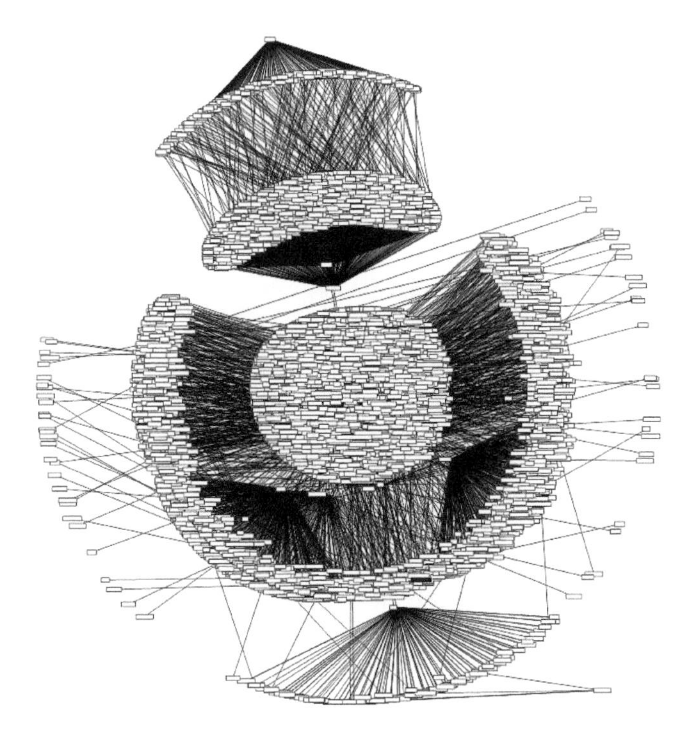

Figure 4.2
A graph generated from file reference statistics for *SM3*

of "parsing" as my informant saw it (detailed in World 2-3): reading data and placing it within an internal representation of the software system. But simply loading the data is not enough to make a game. Nothing has even been displayed on the screen at this point. The loaded data must then be combined, interpreted, and displayed on the screen and respond to the actions of the user.

Engineers are responsible for the tools and workflows that parse and assemble game assets from artists, designers, and other engineers. The creation of these tools is a process, one that demands that an engineer is capable of working closely and collaborating with all the other developers on the team. The tool-building process also frequently requires engineers to understand the foundations of rendering, effects, and cinematic principles that will be underlying the understandings of artists and designers.

The next difficulty for engineers lies in the fact that most videogames do not use a common data format, and this is further complicated on consoles because of proprietary file systems, data formats, and application programming interfaces through which they are accessed, all of which is covered by NDA. Each game engine—the number-crunching and number-parsing heart of a game—makes use of different capabilities from other applications that may be unavailable in others. While things like extensible markup language (XML) or other attempts at industry standard formats have had some traction, frequently these systems are so generalized that they also require a great deal of engine support and specification for them to be of any use. In an effort to maximize use of game media or processing power, engineers will frequently strip unused data from files, leaving data-parsing systems specific to each game.

Because of this, engineers are frequently seen as the gatekeepers to game functionality, as they must implement the systems that expose functionality to artists and designers. Engineers become sentinels because they ultimately have to answer questions regarding hardware resources available to the developers. "No, that will take too much memory . . . Well, if we do that we'll have to rework the way in which those files are being read . . . That will definitely break our CPU budget." One engineer told me that his customer is not the gamer, but rather the artists and designers. He was there to make what they wanted to create functional, to the extent possible considering time and hardware.

This view of engineers as gatekeepers can cause friction among developers. Artists sometimes perceive the unwillingness of an engineer to expose functionality as a lack of interest in the overall visual appeal of a game. Engineers can frequently perceive the demands of artists as superfluous or distracting, frustrating obstacles to the completion of other aspects of game functionality.

World 4-2: Interactive Artistry

It was during production that I really began to appreciate how many artists perceive themselves as those most directly responsible for the creation of the massive amount of content found in modern games. Artists would have lists of the artistic assets they were responsible for creating. Ideally, once created, these assets could then be laid to rest while artists moved to the next item on their list. Unfortunately, artists must often bear the brunt of changes made to the underlying systems of a game: models must be reworked and reexported, textures must be resized and reapplied to models, naming conventions for animations or file structures must be adjusted quickly to accommodate changes made throughout the game's build or tools.

Box 4.2

#: SET DEMO_MODE 1
AUTHOR_DEV_DILEMMA: In some cases, changes must be made for non-technological reasons. The ability to make changes and adjust one's work in ways that can be quickly iterated on is crucial to the process of game development. Artists will often return to a particular asset numerous times through the production process of a game. If the process by which that asset is then made available to the engine or how they are able to examine the results of their work too arduous, in the result is a collaborative disconnect between team members.
Casey: So making that leap from a sketch from design to the art level, that is the hardest part?
ART_DS_Ogre_1: I guess. I realized recently that I was making artwork that didn't really match the style of ART_DS_Ogre_3. It didn't even occur to me to worry about that until the other day. I looked at it and thought, "Oh, wow. His stuff is really cartoony." I thought I was making my stuff cartoony, but comparatively it just isn't so. Now I have to figure out how to make my stuff look like ART_DS_Ogre_3's. I feel like I'm always running into things I hadn't anticipated. I guess I'm just too new.
Casey: But when you have multiple artists with different styles and interpretations, how do you resolve that? Do you do peer review across segments?
ART_DS_Ogre_1: No. Not officially, but we should. But we have been really pressed for time and there are only a couple of us. I guess we just assume we're looking at one another's work and I just didn't think about it.
AUTHOR_DEV_DILEMMA: It is also difficult to often see outside the small production space provided for each individual by the time a team is in production. They are expected to produce a defined set of elements that can then be read into the game and used by designers. Yet, this compartmentalizing, which is precisely what must be done during production, can make it difficult for members of a team to know what one another is doing.
#: SET DEMO_MODE 0

Throughout the production of *SM3*, changes in process and underlying systems required modifications to artwork. In many games' production processes, new tools that artists must be able to quickly grasp and work with are often rolled out during production. In many cases, these custom technologies may not have been designed by artists and an artist might spend several hours attempting to learn how to use the given tool. In addition to tools that change with each project, even well into the project's life, artists also have to work with technology via their teammates. When, for instance, assets that an artist has spent so much time creating do not appear in game or do not look as expected, artists must frequently access the knowledge of engineers and designers. Some *SM3* artists were quick to engage tools engineers, technical artists, and engineers when they were unable to make systems work. In other cases artists would tweak models, animations, or textures attempting to determine why something that ought to be working a particular way wasn't. Their efforts at being technologically self-sufficient were futile, however, for in some cases vague error messages (sometimes displayed by the lovely engineering lead pop-up mentioned previously) would indicate something was wrong with a given art asset and in others applications would simply crash or do nothing.

Functionally, the work of artists differs little from that of engineers, though their tools and knowledge are quite different. In appearance most developers spend the majority of their days seated at a computer or in conversation with other members of the development team. However, the elements that they bring to a game are quite different. Artists create assets that are then often processed by the computer in a way that the images are brought into the game system. Artists will tweak elements of their work to ensure it appears as correctly as possible within the game's systems. An artist will flip between their typical tools (Max/Maya or Photoshop, discussed in more detail below) export a model, animation, or texture, and then view those changes within the game's engine. In some cases there are significant differences between those two views, which the artist must negotiate with or without help.

Modelers and animators work primarily within 3D Studio Max (or simply Max); figure 4.3 is how the Max interface looks when an artist is editing the mesh of a model. The mesh is the collection of lines and vertices (points) that define the structure of the model. In contrast to modelers and animators, texture artists primarily work within Adobe Photoshop, though they continually examine how the changes made in one program affect the other. All of the data they create for these models are stored within a single Max file or texture file. Much like an engineer's IDE, these tools are the

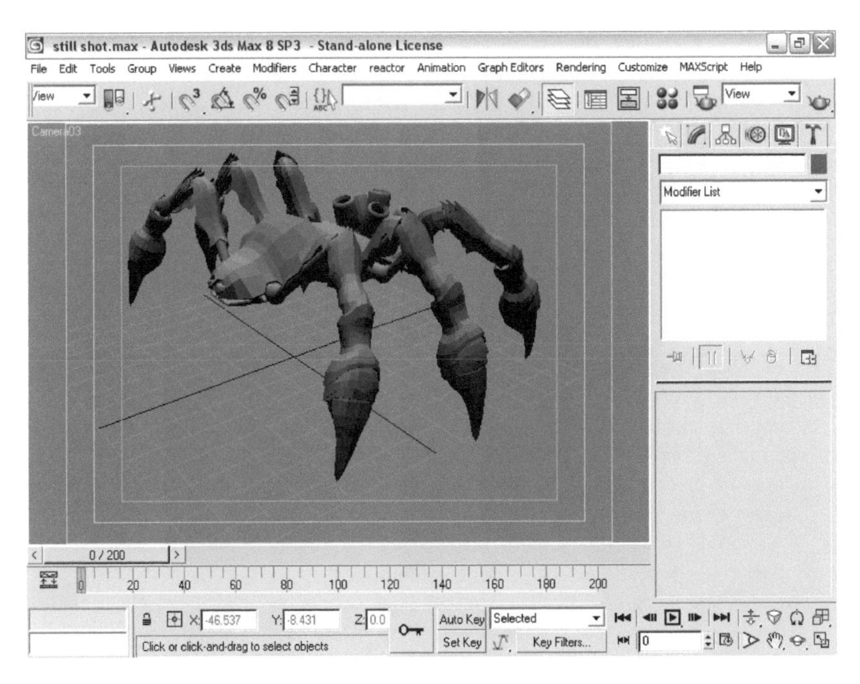

Figure 4.3
A modeler's view of the world in 3D Studio Max

experimental apparatuses within which artists work. And like engineers, artists typically use the VCS to store a historical record of their work. Unlike engineers however, there is no choice in being able to work on a file that has already been checked out. Because of the data formats of these files, artists are unable to simultaneously work on the same files.[2]

Box 4.3

#: SET DEMO_MODE 1
AUTHOR_DEV_DILEMMA: Artists, too, think that the specifics of what goes into their work, though quite visible in a game, are misunderstood to be entirely different. Even when their work draws on, or is part of a broader ecosystem of game production, the work of actually making those things function as a game is difficult to convey clearly.
Casey: How would you go about explaining what you do every day?
ART_DS_Ogre_1: I don't think I understand what I do every day. Really, I've given up on explaining. Most people think that I

(Continued)

```
do work like what comes out of Pixar. The majority of people
only think they comprehend Pixar's animation and that is what
they think I do. They see a 3D animation and they are like, "Oh,
look!" Most people know I work in games, but they have no real
concept of what it is. Some people think I do graphic design.
There is also this misconception that when you're working on a
licensed title that all we do is take someone else's game and
convert it into something that can be played on another platform.
They have no idea that most games are totally recreated. A DS
game may be completely different from its corresponding PC game.
I'm sure some gamers know that, because they play them both and
see that they are different. But so many people think, "Oh, you
just take their art assets and convert them." They assume that
there is no difference across platforms. There is so much design
and creativity that goes into those new games. They are literally
largely created from scratch. We just use the same characters as
a launching point.
   When I first got here I went to ART_DS_Ogre_2 and asked,
"Can't we just take their models and sort of optimize them?" He
said, "It is just faster to make them ourselves really." At least
that way we know ours aren't going to give us any goofy problems
in the tool chain.
AUTHOR_DEV_DILEMMA: This is the work of art game production in
many AAA game studios, where assets may be available from other
projects, but because of the numerous software systems that sur-
round the activity of game development, it is often impossible
for these to flow seamlessly between one project and another.
This complexity of these processes makes it difficult for art-
ists to convey the work of game development properly. Thus, even
though a particular art asset may be visible in the game, the
actual work associated with its creation is quite absent.
#: SET DEMO_MODE 0
```

Most artists on teams in the United States have the opportunity to see their artwork "in game" on a daily if not more frequent basis. They understand why they are limited or "working within budgets" for polygon counts, vertex counts, texture size and shape, and the numerous other expectations that accompany producing artwork for games. If they do not understand the limitations or if they are curious about those limitations (some, for instance, are harder or softer than others), they can ask their lead, or an artist who was part of the team during preproduction when those limits were decided upon, or they can walk over to an engineer and ask. Some will even experiment, simply determining which limits were hard or soft based upon whether those changes break the game or do not.

Box 4.4

#: SET DEMO_MODE 1
AUTHOR_DEV_DILEMMA: Art production is highly contingent on the code that supports it. Changes to the underlying systems of a game make it difficult for art assets to move from one project to another. It also creates the possibility that changes made to the code of a game can result in significant rework on the part of an artist.
Casey: [Laughs]
ART_DS_Ogre_1: I was expecting there to be this massive hard drive with lots of reference material and that there would be established methods of doing things. For example, "This is how we make fire." You know, or whatever else has been done a hundred different times. There isn't much of a shared pipeline. I imagined that there would be somewhere to go look at basic textures.
Casey: So you end up doing a lot from scratch?
ART_DS_Ogre_1: Yes, which is fine. But in a company this big, it is strange, for resources to not be shared. . . . At all.
Casey: How do you like the project so far?
ART_DS_Ogre_1: No complaints. It's awesome. I want to work here forever. There are some things that go on, but I'm just a worker, I don't pay attention. It isn't as if I'm consulted every time some issue comes up. We're trying to do a lot of things on this project that haven't been done before. So, it's a situation where sometimes to reach those new things, we have to do things over again. For example, we recently changed the way levels are constructed. At first it was rather free form and now we've switched to tile sets. Although now we may be switching back to free form. But I had never done tile sets before and free form was very frustrating so I was thinking, "Oh, tile sets, maybe that will be easier." As soon as we were doing tile sets, I was like, "Oh, my god. My head hurts." But those are just my own personal weaknesses.
AUTHOR_DEV_DILEMMA: Many artists talk about the way small changes to a game can result in significant rework and that the day-to-day practices (like making "tiles" or being able to "freeform") can often mean that an artist will have to completely remake a particular art asset as changes are made to the game in order to keep it functioning. Ideally these are issues that would have been determined during the preproduction process, but in many cases developers find themselves for various reason (both legitimate and illegitimate) redesigning basic functionality while a game is in production.
#: SET DEMO_MODE 0

Many of the developers in India with whom I spent time were working on some aspect of artistic production for games being developed by US studios. In most cases the artwork was for established game franchises, sequels in many cases. These games were already in full production mode, with the limitations and requirements for artists already established. Almost none of the artists working with these games saw what they were creating within the game itself until after it had been released. Comments from US-based artists and designers would come back to them annotated for changes, but there was no clear understanding of why or how the limitations that they were working within had been established. Teams of artists were being asked to produce art assets in a way that no traditional game developer would be asked to do. The inability to see how changes made to a particular art asset were then reflected in the games engine made it difficult for artists to understand the relationship between their work and the game in which it was being placed. The interactive and experimental processes were removed and, worse yet, made completely unavailable. This proved particularly difficult for an Indian team of artists and engineers working on a prototype for the Nintendo DS. Without an experience of working with any of the libraries for a console like the DS, the engineers were largely left to learn the conventions of the system from Nintendo-supplied documentation and private messaging boards. These resources did not make it clear how to actually make games in practice, either for the engineer or for the artist. Instead the artists and engineers found themselves walking from desk to desk, transferring files, converting files, asking one another questions, making changes, looking at the debugger together, and struggling with the lack of interactivity.

I was struck at how a team of developers in Chennai, India, were struggling to piece together the foundations of a game's asset pipeline for themselves. My years in the game industry came tumbling back, reminding me that this was a fundamental aspect of game development that was simply not being communicated. Tools and pipelines must be constructed in such a way that each developer can do what they do well, rather than stepping on one another's toes. It seemed as though a lot of talented developers were wasting time and energy doing their work in the dark with their hands tied behind their backs.

As mentioned before, pipelines are the least talked about, least documented, and frequently most critical points in the game development process. And more than any other time, change or redefinition of the pipeline results in massive amounts of rework for artists, engineers, and designers. Though pipelines are frequently defined at the beginning of a project, they must often go through several iterations to adjust to the changing needs of a project. Unfortunately, changes typically occur much later in a game's development cycle than most would like. Each developer experiences only a portion of the pipeline. While an engineer will interact frequently with those components that process code, an artist's experience of the pipeline will often extend into external tools not created by a game development team.

The process of making the transition from the tools of the artist into the game is frequently called the "art pipeline." The artist's pipeline typically begins in Max/Maya or Photoshop. Tools engineers or technical artists often will have customized the extensive set of scripting and exporting features of these programs, which extracts the relevant information for placement into the game engine. For most artists in the United States, this means by the time full blown production on a game has begun, they will have a "make art" button within Max that will export the necessary data from their work in a format such that it can be seen within game. This allows artists to more quickly tweak and view their work. Artists are limited in that there is no debugger for this process. If something does not work or does not appear correctly, there is often no obvious way to determine how or why. Artists typically then proceed to make changes to their models, textures, or animations in the hope of feeling out the reasons. Sometimes they'll turn to an engineer to ask if something can be done or if there is a way to do something that is causing them problems. This is often a point of conflict for artists and engineers, who continue to negotiate pipelines throughout the development of a game. Both artist and engineer have a different conceptual understanding of how something should be done or even what is possible. "Shader," "vertex," "stripping," and numerous other terms mean very different things to artist and programmer.

This is where the experience of the "gate-keeping" is felt explicitly, though tension between specialties is, as we saw in World 3, broadly labeled as a "communication issue" in studios. Because technical artists and tools engineers are responsible for the creation of the tools that power these pipelines, they are also often responsible for other things, and time is always at a premium, causing conflict to frequently arise. One informant and I attempted to spell out the enduring tensions of pipeline clashes in a GDC session presentation proposal:

> Artists and programmers have worked together in games ever since the first game programmers said to themselves, "My art sucks." From that day forward, we have tried to integrate artists and their craft into this highly technical field. Here in 2006, we should consider this a work in progress that all the disciplines of game

development can endeavor to improve upon. This session dissects common issues and provides solutions in the artist/programmer relationship that development teams of all sizes face.

A few months ago, the two presenters spent time speaking with both programmers and artists at Vicarious Visions. They conducted a one-hour roundtable session for artists only, where they could talk about what they did and did not understand about programmers. Then they ran the same session with only the engineers. The one thing that amazed both of the presenters was how professional and genuine both sides were. They both wanted only the best for the game and their team. How then, could they end up at each other's throats in the middle of development? What is the problem?

The presenters will begin by having the audience ask themselves the following questions:

Artists:

• Have you ever tried to suggest a feature to a programmer, only to walk away frustrated and upset?

• Are the in-house tools that you use bug-ridden and overly complex?

• When you have a problem that programmers can solve, do you hesitate to ask for fear of the response?

• Do you find yourself overwhelmed with techno-babble?

• Do you ever feel cut out of the loop in designing your own workflow?

• Do you ever operate outside of your team structure and go to a programmer on another team for advice?

Programmers:

• Do artists break your tools and game code with stunning regularity?

• Have you ever given an artist a checklist of steps to follow, only to have them fail to do so repeatedly?

• Have you found artists performing mind-numbingly repetitive tasks that you could have fixed with code had you only known about them?

• Have you ever needed a simple art fix, only to have an artist tell you that you are asking for the impossible? (Informant and O'Donnell 2005b)

Videogame artistry, in many respects, is interactive based upon the tools that define it. Yet that interactivity can break down, resulting in a different kind of interactivity, one in which artists and engineers can "end up at each other's throats." In many respects I see this breakdown related to "standards" and "classifications" that "may become more visible, especially when they break down or become objects of contention" (Bowker and Star 1999, 2–3). But it goes beyond just social relations or disciplinary differences, or differentials of power in the setting of standards and classifications, which are typically developed in already compacted inconvenient time frames. It has just as much to do with "institutional" deficiencies that prevent broader discourse about "standards" or "classifications." The demands for secrecy throughout the industry prevent that. Thus, the breakdown of interactivity is partially rooted in disciplinary ways of understanding how things function and what is being created. It is also rooted in a kind of institutional Alzheimer's that continues to be perpetuated through the very structures of secrecy that lend the game industry its panache. It is based, too, in lack of conventions, standards, documentation, public discourse, and institutions. And it can all be felt even more explicitly in the everyday lives of game designers.

World 4-3: Designing Interactivity Interactively

The role of game designer as singular, specified, and locatable in videogame organizations and educational programs has been relatively recent. In previous generations of game development, having designer/artists, designer/producer or designer/engineers handling the design of a game's mechanics or systems was much more frequently the norm.

Box 4.5

#: SET DEMO_MODE 1

AUTHOR_DEV_DILEMMA: Very few designers who I spoke with in the game industry were "trained" designers. Most came from a variety of disciplines and somehow wound up, by either desire or accident in the role of the designer. Most found the process of design interesting and had some previous experience designing either as a hobby or recreationally games, but very few had a specific degree in game design.

Casey: Let's start at the beginning. Give me a little history of how you wound up at VV and found your way to now.

DESIGN_LEAD_1: Well, I'll start back at the beginning, when I was a little child. . . . No. I guess it started my freshman year of college. I lived with a bunch of guys and a couple of us became very close friends. That was the first year that I had a computer that I could call my own. It was also the year that DOOM came out. So a couple of us started making DOOM levels, you know, back when the editing software was a pile of crap. But, I was good at texture alignment; I could just look at it and guess which pixels were off. Anyway, we started making levels and working together and ultimately I became engrossed in videogames throughout college. After we finished, we went to work for MGMT_CREATIVE_HEAD_1. I saw some of the work they were doing and said I thought

(Continued)

I could do better. He said, "Show me." Eventually they hired me. The rationale being that I might as well do it while I was young. It was really a chance to do something I never really imagined myself doing.

There was no sense of a "design" role at all at VV at that point. Everyone chipped in and did design for things they were interested in. I ended up doing mostly resource economies and narrative, because I guess someone thought I could write a story, or I convinced them that I could even though I had no clue. Probably because I was a film buff more than any other reason. Resource economies because of all the MUDs and online multi-player games I had played. Eventually we just formed a design role out of that. Then we start hiring more people and define the position and at a certain point you realize that you need some type of cohesive group. So several other designers and I formed the design group, but I didn't want to manage it, so we hired MGMT_DESIGN_HEAD_1 to do that and built the infrastructure around it.

Casey: So, given the amount of time you've been with the company, do you feel like you're still proving yourself? Or do you have that freedom to just say, "Trust me?"

DESIGN_LEAD_1: Yeah. You need a culture that allows people to make mistakes. That is something that I've been trying to push. If you don't strive to innovate, you never will. You are just going to try and avoid failure at that point, and avoiding failure is not how you're going to make successful games. You can make a mediocre game that may sell well, but that isn't going to keep your team around and that isn't going to make you feel creatively fulfilled. And, when you're constantly doing due-diligence and trying to prove your points, you are focusing more on avoiding failure than trying to succeed. So I've really picked up the mantra, "fail quickly." But, you have to be able to fail, to fail quickly.

AUTHOR_DEV_DILEMMA: Design, more than other aspect of the game development process was the most open and subject to change throughout the development of a game. For some designers it was a process of "finding the fun," which was often put in tension with other requirements of a project. Always it was an iterative process that had far-reaching consequences for teams in production. As a game is developed new avenues are explored, and often when designers begin to find new "fun" in a game, pursuing those features could require significant shift in a game's production.

#: SET DEMO_MODE 0

During production, designers are responsible for putting many of the pieces of a game together. While the basic systems and mechanics of a game will have been defined during preproduction, during production all of the specific elements need to be assembled into a game. Levels need to be created and balanced. Missions need to be defined and tested for difficulty and plausibility within the storyline of the game. Specific game objects need to be placed or their behavior scripted in such a way to make it specific to a time, event, or other element. All of this information is referred to, vaguely at best, as "data." Data forms the glue between the game's engine created by the engineering team and the art assets generated by artists. Engineering defines the structure by which the game can be specified, artists create visual elements that can be pulled into the game, and the designers create data that draws on artistic assets and places it into the structure indicated by the engineering team.

For the most part, generating data consists of creating files, which are combined with artistic "assets" and interpreted by the underlying source code written by engineers. This underlying code can be an XML file or files containing scripting languages like C#, Ruby, or LUA that direct the game-code, telling it how to behave. Scripting is similar to programming, however scripts are interpreted during the execution of a game rather than being precompiled into the native machine code. This combination of art and data passed through the underlying code defines the structure of the game. Sometimes these coding activities are enabled and assisted by custom tools built by tools engineers and technical artists at their company, other times external software packages are purchased, and other times they may only be able to work with a text editor. This more primitive option can intensify the feedback loops among engineers, artists, sound engineers, and designers. If changes must be made in different places to accommodate new concepts or approaches as a game is developed, then changes may be required throughout the different components that make up a game. This is not the "fault" of designers, but rather a product of their position in the creation process. This means that often designers find themselves sitting in front of text editors or spreadsheets thinking about the economics or relationships between the different components of a game's underlying systems, creating scenarios, or scripting parts of a game to ensure that particular storylines are delivered to the player or how aspects of the game will react differently to a player based on the players activities. In some cases, custom tools are created for designers that facilitate these activities. For small game development teams, or those early in development (prior to the stage when custom tools are created) designers are primarily defining data in some way that the game can easily read it. XML is a popular format for structuring data and more easily parsed by the underlying code. In other cases designers will work with various scripting languages that allow them to articulate, in a fashion understandable to the game's engine, how objects relate, interact or change the game's underlying state. Thus, for many designers, the text editor is one of their most important tools.

Designers occasionally mix up this standard amalgamation of computer time by playing games that have preceded the one they are currently working on, with a critical eye toward what is enjoyable, and what is not. This is likely where the fallacy that game developers "play games all day," derives from. However, most game designers play games in a highly critical mode, dissecting the elements, namely, "Oh, there are the '*God of War*' style quick-time events," or "Uh, oh, skill tree ahead!" Other games are played to work out new game mechanics, which might become useful in the development of their own components.

Perhaps second only to tools engineers and technical artists, designers must have excellent communications skills. They must be able to collaborate and work well with one another, as well as the engineers and artists, which they connect through their rather abstract job of generating data for a game's engine.

Designers were the most difficult game developers to find in India, in part because designers are almost always gamers and in India gamers can be difficult to find. Beyond the general derision that parents level at videogames the idea that game development is a profession has not yet caught hold. A son or daughter interested in making games faces the assumption by parents that they "are going to be playing games all day." Videogames are still viewed as a diversion from those educational tasks that students ought to be preparing for. For this reason, design has been a difficult leap in the Indian industry. This is made more problematic by the general lack of professionalization of game design or game development more broadly. The technical, social, and procedural connection of engineers, artists, and designers in ways that enable collaboration has also been difficult. Because these practices have been developed experimentally over time, through experience, and are entirely undocumented, means that they are rarely communicated outside studios that develop them.

"Tools," have been essential for the professionalization of designers within game companies. Tools provide designers with structure through which they can help bring together the work of engineers and artists in a meaningful way. However, as discussed in World 3, most tools are project or team or company specific, making it critically difficult for designers to

Box 4.6

#: SET DEMO_MODE 1
AUTHOR_DEV_DILEMMA: Tools engineers are those that focus specifically on software systems that connect artists, designers, and the engineers working on a game's engine. Tools can also provide other support mechanisms to a team. Build tools, testing tools, etc., all fall under the purview of the tools engineer. They are a particular breed of engineer, because they often do not find themselves in the limelight, though players may actually use their tools if they are released to users in order to enable MODs.
Casey: So, you've always been interested in tools even while in undergrad?
TOOLS_ENG_2: Yes. My thesis was really more of a user interface sort of concept. The idea was to take some of the complex work of 3D modeling in Maya and Max and simplify it. Each application has all of these different operations for how you interact with a model; you can grab a couple, or a group of vertices and pull them this way or that, but it can be really tedious because you can have thousands of vertices that are visible. So we were trying to make the whole system more gesture based or sketch based. You can take your mouse and sketch what you wanted to do rather than the old method.
Casey: Now you are part of the *SM3* team, what is your official role?
TOOLS_ENG_2: My title is associate tools programmer, so my role is basically working with TOOLS_ENG_LEAD, for the most part. I'm spending most of my time on Peaches and other *SM3* tools.
Casey: How is that going?
TOOLS_ENG_2: It's going well. I think it's going well. Peaches is coming along. The hardest part for me is that I feel like, and this could just be a personal thing, that with all of the things we have to do, I feel like things could be getting done faster or more efficiently. I'm concerned about time constraints but no one has said if I should be doing things faster or if turnaround is good enough, so it is hard to gauge. There is also the scale of the project, it's hard to keep up with what is going on sometimes, outside our little tools world.
Casey: Yeah, they have nine engineers, or so, working on all different components.
TOOLS_ENG_2: Because of that, a lot of times I kind of have to look to TOOLS_ENG_LEAD to help buffer. Even within Peaches, it is such a big program, I will start working on a tool and have to keep asking him, "Where is this thing?" I don't know. There is a huge code base and I don't understand how it interacts with the game fully. Yet.

(*Continued*)

```
Casey: And you have things like Alchemy too . . .
TOOLS_ENG_2: Yeah. I feel a lot better about that now, but at
first I was like, "What is an IGB file? What are you talking
about?" It was definitely intimidating at first. The cool thing
about tools is that every time you work with a tool, you get to
touch a different part of it. So I learn as I go.
AUTHOR_DEV_DILEMMA: Peaches was a data editor first created spe-
cifically for the SM3 project. Alchemy was a set of tools that
defined a kind of basic tech upon which other games could be
created. Tools touch every aspect of a game. They are part of an
increasing desire for game development practice to be less cha-
otic and developers to specialize. Tools are thus about attempt-
ing to solidify the practice of what it means to make games.
#: SET DEMO_MODE 0
```

professionalize outside the industry. Thus, the mantra for those hoping to be game designers—"go make games." This is of course correct, but the kind of literature that helps define what that means for a designer is not so simply available.

As previously mentioned, there are no commercially available "standard" tools for designers to use in the process of making games. Instead, the job at each company of technical artists and tools engineers is to create new software systems for designers, artists, and engineers. While there are numerous "middleware" companies creating tools, frequently these fill engine gaps rather than tools gaps, meaning that they are designed to be used in the underlying "tech" or code of a game, and while designers have tools of their own, these tools cannot link together all of the other pieces of design as must be done to create a game. That is the job of the engineer.

Tech is one of the most important pieces of code in game companies: the foundational pieces that form the core of every engine or game that a team of developers creates. Tech can include aspects of the pipelines as well. My primary field site, in particular, spent years and massive amounts of money and mind power to create their internal tech. This system was originally purchased from a company called Alchemy, which retains the name. More recently, the goal has become grander than simply a game engine; the company now wants to create a foundational layer of tech and tools that can support not only a standard art pipeline, but also the data pipeline of designers.[3] It was with this goal in mind that Peaches was created. The name was based on a historical practice of the tools team that stipulated that all tools would be named after some sort of food, which could later be justified by an accompanying acronym.

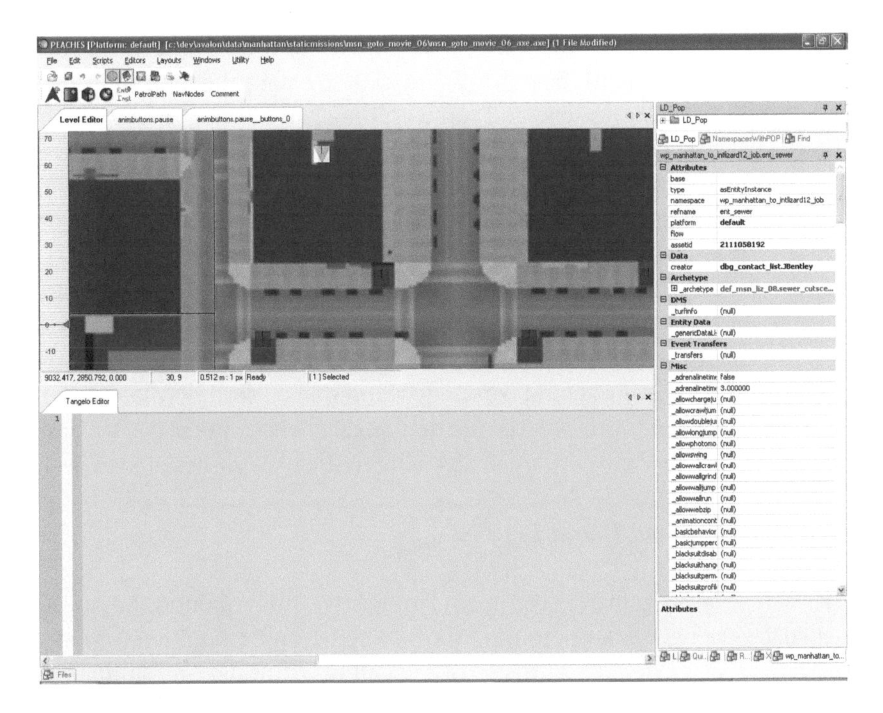

Figure 4.4
Peaches in action editing a game level

Peaches, seen in figure 4.4, was created in part to assist in dealing with the complexity of a project like *SM3* (the vastness of which figure 4.2 illustrates). The sheer number of files and references between files requires a new set of tools for developers. While level editors and other systems have been quite common in the past among game development companies, the goal of Peaches was much broader. It was designed to be a system that could be expanded on, as new kinds of design data were necessary in future projects. The same tool could be used by designers to create special effects, levels, and missions and to script cut scenes.

Even when working independently with tools, though, the activities of game designers are intertwined with those of engineers and artists. Each depends on the other for the successful completion of a game. In the interim, with no standards and no standardized tools, designers must constantly work with and without artists and engineers in the construction of virtual spaces, story-lines, and characters for which there is no agreed upon language. Designers, much like videogame artists, are constrained by their tools, the game's tech as defined by engineers, and the ability to translate abstract concepts into forms that work with each of these. Designers do not direct their ire at engineers, however. Instead, they take the tools they have managed to cobble together, and run with them, attempting to make them do things they were never intended to do, in the hopes that what results is indeed "fun."

If designers focus their frustrations in any single location, it is only on the constraints driven by time, and on the frequent inability to return to decisions made previously. Designers forge ahead because the game is ultimately their vision with whatever tools they find at hand. The designer is the exemplary bricoleur, attempting to create spaces of play amid a myriad of artistic, technological, and corporate restrictions placed on them. Yet, it is the set of systems that they create that largely defines the player's world, though their labor may be largely invisible to the player when done well.

World 4-4: Keeping Things Synched

At each level of game development, interactive tools abound. Engineers use IDEs, debuggers, and profiling tools to better understand the flow and execution of game code. Artists use modeling, painting, and animation tools to craft the visuals to be displayed in the game. Designers often use custom tools to combine art assets into game spaces governed by a combination of game code and rules, missions, and stories designed to engage the player. All of these interactive tools must have pipelines that allow data to flow from one discipline to another. This is where systems begin to break down, in part because the pipeline occurs at precisely those cross-disciplinary boundaries that have made the emergence of technical artists and tools engineers necessary. The pipeline also fractures most spectacularly because it is the most subject to breakage when fault lines move suddenly.

When developers talk about what is missing, about the aspect of game development that prevents them from being able to work well, they frequently settle on the highly problematic term "vision." As they conceptualize it, vision is a clear idea of what you want at the end of a project. Vision is assumed to come (or not come) from somewhere above in the company, delivered by management to help developers understand how to direct their experimental efforts. When a vision is combined with a plan for how that vision can be brought to life, the work can then be scheduled. Unfortunately because of the constantly changing technological landscape, not only is the vision often missing, but the subsequent "how to implement it" is often absent. But the designers' lament for a nonexistent vision "ignores the fact that outcomes are socially accomplished in context rather than individually calculated . . . it ignores the fact that outcomes are often not consciously calculated, or even intended by any one of the parties involved" (Knorr-Cetina 1983, 130). Vision is accomplished along the way. Designers thus often desire the very thing that they cannot have. Yet in a work environment where predetermined end products and paths toward them cannot be known, the ability to plan is significantly compromised. Without plans for the *what* and *how*, designers balk. The conversation below indicates how many developers have come to see vision as central to their undertakings.

Casey: So, what do you think went wrong?

ENG_LEAD_Asylum: Well, if you want my opinion . . . , it's lack of vision. If you know what you want, then you can get it sooner, but most of the time, you don't know what you want, so you have to see it. At the beginning, you don't really know what you want, so you just kind of try things out, and you get one thing, but you can't really tell how that works, because you need these other things in place, but you didn't know quite what the other things would be, because there isn't an overall . . . theme. And then, the later you go, the more concrete things get, and its more apparent with the pieces that you have, what you have to do, and so then you end up with something. At preproduction, everything is very free, flowing, and then when you go into production, and try to execute, and then you're at the end of production, and you see how everything turned out, which is very different from where you started. But if you had a clear vision of what you wanted at the start, then things might have turned out differently. (Informant and O'Donnell 2005e)

Having spent three years observing (and more than seven working) in game studios in the United States and India, I tend to agree that games, which begin without a clear vision of what they are supposed to be, have trouble being implemented prior to their deadlines. At the same time that vision shifts over time during the development process become more refined as it is explored. Furthermore, games without a clear sense of purpose are also those that fall prey to the multifarious desires of the companies that fund its creation. Designers, more than any single category of developer, are responsible for defining this vision. This is part documentation (writing down what the game should be), but it is also about thinking about how all the pieces should relate to one another. Undoubtedly, they will shift as a game is developed. But without an objective, many games wander aimlessly into production. Without a clear vision, there is no reason to protest "feature creep" (an industry pejorative for the additional elements that simply have no place in the game being developed). Developers often run after half-formed or vague ideas for additional features, only to find themselves miles down a road that was not the best to first travel.[4]

Perhaps it is too idealized to expect to know precisely what you want and how you are going to get it. After all, we've come a long way since the "waterfall" method of design concept from the dark ages of game or software development. The waterfall method was based on the idea that a team would know explicitly all of the requirements of a project before setting out to build it; however, as already noted, game development is much more experimental, requiring changes to the requirements throughout the process. Further, developers must always answer to someone up the food chain and be capable of arguing for or against changes that may very well derail a project. My informants and I seek a clear vision of the final product with the acknowledgment that the target will likely shift in new directions throughout the life of a project. The distinction is that the game will swerve in directions that fit the overall vision, rather than turning in random directions that allow the overall structure and rules to take control of the development process.

This vision problem becomes particularly daunting for Indian developers. While they frequently have more detailed contracts that govern their relationships with developers in the United States, the constant flux of project needs in one location impacts its partner site. For example, one location will have to scramble to rework art assets when the art pipeline changes in a mirror location. Managers must either negotiate change orders with the studio that has contracted them, or have employees make the changes without adjustment to the contract. Because US developers are so accustomed to rework, Indian studios risk amiable social relations when they ask for compensation for rework activities, which US developers assume are "natural." Because complete overhauls to the work in the United States and abroad strain workers, budgets, and timelines, lack of vision means more work, more tension, and more conflict.

Vision has really become shorthand for the goal that crosses each of the disciplinary boundaries within a game. It represents the idealized notion that having a common goal in mind will help keep these systems in sync across their numerous fault lines. But even by game developers' own admissions, oftentimes the idea of what a game should be at the beginning is quite different from what it is at the end (Hoffman 2009). Even when one turns to a more historical view of scientific and technological production, it seems difficult to privilege the vision, as it will shift over the life of a particular project. The "mangled" (Pickering 1995) character of technological development lends itself to thinking in terms of "resistance" and "accommodation" (Pickering 1995). The privileging of vision is thus a kind of technological determinist position, though rather than privileging the technological it privileges the vision. Thus, as Knorr-Cetina notes above, a "vision" is kind of a false hope for game developers. Design itself is more frequently thought to be a dynamic process that involves numerous points of experimentation and trial and error (Bucciarelli and Kuhn 1997). What game developers are ultimately seeking is a means to improve the flow of information and communication across fault lines. Because, ultimately, the "vision" of a game is caught up in the ability to keep that vision communicated in a way that speaks to all of the disciplines involved in its production. If the vision of what a game "should be" is both shifting and experimentally determined, then ensuring that that vision can be updated and communicated quickly and accurately are core to the problems facing developers. At the same time, because organizations frequently believe they cannot spare enough engineers and artists to fill the roles that attempt to bridge fault lines, game development organizations find themselves grappling with stressed systems that ultimately fail when in full production mode.

It is important to point out that when these highly coupled systems break, they break spectacularly, and do so more frequently when they have no documentation.[5] They are highly coupled in both a technological sense and human sense. Numerous technologies touch one another and numerous disciplinary understandings are also interconnected throughout the system. Implicit assumptions based on previous experience may prove more volatile than expected because of the shifting context of work.

Because of the sheer number of disciplines, technologies, and practices that make up the practice of game development work, there are a large number of interconnected and dependent pieces. As we've seen before, artists or designers cannot see the results of their work until engineers create the underlying code or tech to support those features. Artists may be unable to see their work in a game without data defined by designers. And, designers may not be able to see the results of their work without the associated art assets of the artists. At the same time, numerous complex software systems are mediating the interaction between these individuals. The build system itself is a system that may fail or break, regardless of the health of the underlying game. In other cases, a game may function on one system, but not another.[6]

While "incidents are overwhelmingly the most common untoward system events," I suspect that given the frenetic pace of game development, lack of broader discussion of best practices, or any practices for that matter, the commonality of full-fledged "accidents" becomes much more common. As other industries privatize and place more emphasis on secrecy rather than open discussion, the implications outside of the game industry are troubling. While a full-fledged "system accident" for a group of game developers results in long hours and a stressful work environment, the implications for all creative collaborative workers, whose developments are not so perfectly contained, becomes particularly troubling (Perrow 1999, 70–71).

The general unpredictability and instability of production systems has led many to see the answer in more real time feedback throughout the systems, including the human component. The game development trend has been to move toward what I have termed an "interactive" model of game development, where changes and modifications to the overall complex system can be viewed in real time and instantly.[7] While in some respects, the goal of instant feedback and response can indeed be a boon for developers (as in the ability for an artist to know precisely how and why things are not working as they had thought they might), the same goal does not necessarily extend itself to the realm of human work or work organization. People simply do not move, think, or understand complex problems instantaneously. We need moments of rest, separation, or thought to reach conclusions. These goals surrounding interactivity can simultaneously be overextended, resulting in what I term generically "churn," or the inability for workers to find a reasonable space of time to sit and work on their assignments. Feedback and information for the sake of feedback and information result in situations where systems come to a dynamic standstill.

Creative, collaborative, and interdisciplinary work is difficult already. There is a reason that disastrous portends like "fault lines," "sedimentation," and "volatility" are used in these contexts (Traweek 2000; Fortun 2006). The development process is fraught with the continual (re)formation of creole languages and the experimental process. Tools break and complex systems fail all throughout the development process. In an effort to increase efficiency, interactive systems are deployed, but can distract us or become goals in and of themselves. The conversation between an individual's work process and broader systems is a critical component of where we find meaning. Quite literally, it is the "meaningful play" (Salen and Zimmerman 2004) of work itself. Game developers by and large have been lost in their ability to really reflect, document, or talk about those experiences that would historicize their activities.

World 4 Boss Fight: This Ain't Anything like Grandma's Boy

The ability to interactively work across disciplines to solve problems was almost universally cited by informants as a "useful" or "necessary" aspect of game development. Fast cycle times and feedback loops that allowed small groups of developers to rapidly find solutions to problems, which they faced in the development of a game, were found productive. The project lead of *Bioshock*, again, notes how interdisciplinary collaborative teams were crucially important throughout the development process: "Over the course of development, we created multidisciplinary strike teams to work on a wide variety of problems, including AI, animation, visual effects, and cinematic. The results of those teams were universally better than the previous non-iterative process" (Finley 2007, 24).

At the same time, this process, when under pressure can result in reckless cycling, or rapid changes that result in a more chaotic structure. The ability to "iterate" on a problem with a team is productive, but when that iterative structure is put under immense time pressure, it often begins to fall apart, rendering the resulting incessant iteration useless.

As noted in my fieldwork, unpredictable results may occur when iteration and fast feedback loops are bypassed or disconnected from those structures that attempt to keep them under control. An engineer working on Microsoft's game *Age of Empires*, developed by Ensemble Studios commented on a phenomenon very similar to experiences of my informants during periods of intense activity.

> The lead is the go-to person when someone outside has new requests for the team. As the development of AOE progressed and the pressures rose, adherence to this system broke down as people went direct to get their needs filled quickly. We paid a price for it. People didn't know about programming changes or new art that was added to the game, and the level of confusion rose, creating a time drain and distraction. We all had to stop at times to figure out what was going on. (Pritchard 2003)

System failure can result when interactive interdisciplinarity disconnects from the safety valves between feedback loops. Emergent forms of structure within these groups must be considered as important as the more rigid and formalized structures to allow for the kind of experimental outcomes that form the foundation of effective game development practice. These structures must also be communicated among team members, a process that is frequently neglected when time pressures are imposed, are disconnected from knowledge about where a team is, or are oblivious to how a team is functioning.

Systems are doomed to fail when a small number of individuals working at the margins of disciplines—the tools engineers, technical artists, and leads—are not provided with the human and time resources necessary to prevent breakdowns between groups. Again, the project manager from Insomniac Games comments about the tendency toward excess these systems can exhibit: "Adding to the confusion, only a small number of programmers had the knowledge required to debug the problems, and these people were overwhelmed with requests for help. If it weren't for their inhuman effort and long hours hunched over keyboards, we would have never hit launch date" (Smith 2007, 35).

Unfortunately this bottleneck effect is frequently solved by asking or assuming that employees will stay late to make up the slack created by overwhelmed feedback loops, ineffective interactivity, interdisciplinary breakdowns, or disrespect of emergent forms of structure. Time spent maintaining effective systems is seen as separate from the actual work of making a game. This can no longer be the case if the work of game development really does include this interdisciplinary work; for teams to work together, studios must take into account the structure and process of game development.

For game development to truly find stability or sustainability, the industry must adopt a modicum of standards around which tools and practices can be instituted. Most importantly, these standards, tools and practices need to be open for debate and conversation. They are ultimately too important to be hidden behind NDAs or licenses available only to a small portion of those who call themselves game developers.

World 5: Leeroy Jenkins, Autoplay, and Crunch

Box 5.1

#: SET DEMO_MODE 1
AUTHOR_DEV_DILEMMA: Work/home balance is an issue for every developer who has spent a significant time in the industry. Yet, it seldom remains a consistent issue that developers devote themselves to solving. Game developers struggle with the immateriality of their work, in that it can make it difficult to justify to significant others left to late nights on their own, what precisely they were doing all day. Of course it doesn't help that there may very well be the assumption, though false, that developers might just be playing games all day.
Casey: Has working in games had an effect on your personal life?
ART_Spidey_2: It definitely hasn't helped it. One of the hardest things is how I sit here all day working, but when the computer turns off, I don't really have anything to show for it. I mean, when you're working for eighteen hours and you go home, you can't really show someone all that you did for that length of time. The other part is that it's such a young industry. There are lots of people fresh out of college, who aren't married and love working on games. So it isn't a normal office where people come in at 9 and leave at 5 and if you stay until 8, people are appalled. Here, many people are like, "I got nothing else to do." So if you're the one with a wife at home, it can be hard. But things are flexible, which is good. But probably more bad than good.
AUTHOR_DEV_DILEMMA: So, many game developers dive into the game industry. They start their careers passionate and excited about making games. Many, within a very short period of time, begin to develop a jaded view of the industry. Sometimes it seems as if being jaded about the industry is a kind of badge representing one's actual experience working in it. So many love it, in spite of it. Love and hate mix into a kind of abusive relationship that

(*Continued*)

results in the majority spiraling away from the game industry within five years or less.

Casey: So, do you still love it?

DESIGN_LEAD_1: Did I ever love it?

Casey: That's a good question.

DESIGN_LEAD_1: It is one of those things. Sometimes you love it and sometimes you hate it. I haven't completely lost the romantic notion of game development, because everyone has hopes that they're going to work on something really cool and it's going to do well. Each project that you work on kind of grinds away on that dream, if it's not successful.

I wouldn't want to start my own thing. I know how much work that is. I guess I know what I'm looking for if I ever decide to go somewhere else. I was virtually ground level this time along, and doing that all over again doesn't seem appealing. But there is the quality of life thing too. Developers don't last more than ten years and I've been doing this for ten years. So, now I'm asking, can I keep this up if I wanted a family, much less a girlfriend? The quality of life does suffer. Some people are fine with that. Because they're hermits or programmers or whatever.

At the same time, I can't really go back. Can't do biomedical engineering without going back to school, it's been ten years. I guess I could manage just about any team working on some project. For programmers and artists, when they leave the industry, they have a path. Designers. . . we're kinda screwed. What a lot of people do is start side projects. Things they have control over. I'm going to do more of that. I haven't in the past because I'm just drained every day.

AUTHOR_DEV_DILEMMA: With the emphasis placed so often on "passion," it only makes sense that for so many developers when they walk away, they are really done, their experiences and expertise lost to a "romantic" relationship that simply wasn't balanced to begin with.

Casey: What do you like the least?

TECH_ART_1: Oh, it has to be the culture of overtime. I mean, there is a lot of history in the game industry as rooted in nerds in garages and doing it for the love of it. That has just filtered down into places expecting employees to just go above and beyond without question, all the time. As I've gotten older, I've found that there are other parts of my life that I care about. When I was young, I didn't. I would work a ninety-hour week and be happy. I don't really want to do that anymore. It doesn't mean I care any less about making great games or my career or anything. I think that is a part of the industry that needs to be fixed. It has to be a maturity issue.

Casey: What do you mean "culture of overtime"?

```
TECH_ART_1: I think it has to do with the hyper-competitiveness
in the industry. For people to stay ahead, people making deci-
sions have to push their employees. Often times it's easier to
say, "You. You work. You employees work harder," rather than
having to spend money in areas or hire another person to do
something.
AUTHOR_DEV_DILEMMA: This "work harder" mentality owes debt to the
mythology of the technology startup. The culture of having come
up from a garage persists, and even though many of these compa-
nies have matured, their day-to-day work practices have not. The
culture of overtime authorized in the startup is maintained (even
in consolidated) in multi-national game development publishing
companies where, out of fear of having to break into the game
industry again, many developers work harder rather than pushing
back against management or licensors.
  Perhaps all of this isn't unique to the game industry. And
maybe, game developers might want to consider their work as yet
another creative industry enmeshed in a broad system of global
political-economic networks that shapes their daily work experi-
ence rather than as completely distinct from broader work settings.
#: SET DEMO_MODE 0
```

World 5-1: Managing Chaos

What is it precisely that drives game developers to do what they do, and to do it so intensely in many cases? As we have seen in the World 1 discussion of endemic crunch problems, developers tend to work hard and stay late, encouraged by an industry that chronically mismanages deadlines and demands inhuman work hours for long periods of crunch during production. Most troubling of this intense work environment is the collapse of desire, work, and play into AutoPlay. AutoPlay, a term coined by an anthropologist studying casino game players, "marks the point at which the varied, complex forms of interactivity and productivity that have become the trademark of the 'digital age' loop into recursive forms of disengagement . . . Players cease to be desiring subjects" (Schüll 2005, 78). In other words, AutoPlay marks the transition at which aspects of work/play that encourage our involvement or enjoyment ("fun") in work practice—collapse and disengage—crunch.

In an environment flush with both productive and playful distractions, developers often find themselves much like Schüll's gaming machine players, searching out isolated locations where the pursuit of creation can take place uninterrupted (Schüll 2005, 73). They, too, desire the deep hack mode[1] where lines of code, level design, mission scripts, animation frames, texture art, or model geometry can be produced. The drive to pursue these longings often pushes developers toward excess.

In the spring of 2005, a video clip released on the newly introduced Google Video service began making the rounds of VV's offices. Before the eventual point when everyone had loaded up the video on their own machine, small groups of developers would crowd around a computer watching the clip together, bursting into laughter at one key moment.

The two and a half minute clip was made from within the game *World of Warcraft (WoW)*. It features what appears to be a *WoW* guild, "PALS FOR LIFE," preparing for a "raid" of difficult sections of the game. The team members are busy audio chatting with one another about the tactics and actions they will deploy upon entry into the room. The conversation occupies the first nearly minute and a half of the clip. Their careful preparations are suddenly interrupted when one of the guild members, "Leeroy," returns from being "AFK" or Away from Keyboard only to yell, "Alright, time's up, let's do this! Leeeroooooy Jeeennkins!" Each time the huddled groups of developers watched the moment "Leeroy" charged into battle (to the dismay of his clan brothers, who were discussing the tactics they would so delicately deploy to ensure victory), VV employees would laugh uproariously. In the video, there is a subsequent moment of stunned silence before the raid leader says, "Oh, my god, he just ran in." The clan then attempts to run in and complete the mission only to be "wiped." The video ends with the death of the guild members and much audio chatter about how stupid Leeroy is, to which he replies, "At least I have chicken," which can only be assumed to be his reason for being AFK.[2]

I had forgotten Leeroy for a while, but when I least expected it, it resurfaced. I began to think of the clip as an effective analogy for how, despite all attempts to otherwise stave off defeat, when your "chances of survival" are only "32.33 percent repeating,"[3] things often do not go quite as planned. Complex in different ways than making games, the humorous *WoW* guild raid video clip bears many similarities to the work spaces of game developers. Frequently, despite all of the planning and attempts to manage the process of creating games, game development results in a final melee that bears little resemblance to what many hoped the final battle would look like.

Most games are believed to go through an idealized waterfall process from preproduction to production to testing (or Q/A) to golden master when the final version of a game is sent to the publisher. This idealized process barely scratches the surface of how games actually get developed, but the widely held belief that this is how games are developed means that the notion persists.[4] Even as an idea, however, it so grossly misunderstands the process of game development, which is much more iterative and messy, that it actually harms the industry to understand this as the way games get created. It even goes so far as actively ignoring such well known software engineering ideas like "the mythical man month," and the inherent problematic character of creative technical work (Brooks 1995). This approach ignores assumption in Brooks' argument, a conjecture game developers frequently echo, that until you begin making the thing (in this case a game), it is just an idea that you likely fully don't understand (Hoffman 2009). Perhaps most frightening is how isolated the conversations about the messiness of game development are. Many game companies have attempted to get better at the process by having process managers or people who make it their job to better understand how to more effectively make games.

More recently, many game development studios have begun playing with more broadly establishing software development processes. "Agile" development or one of its incarnations, "Scrum," have been widely touted as making significant improvements to the game development process. Indian game companies in particular have made extensive attempts to bring proven software development best practices into the context of game development. In many cases the upper management of these companies come from other areas of software development where similar methods have improved the management of software production. The steadily growing number of sessions at the Game Developers Conference that feature the words "process" or "management" in their title will give you an idea of the growing popularity of this new area.[5] Some studios in the United States and Western Europe, however, are fighting this interest, asserting the widely held belief that game development is just different, unmanageable. Sometimes, teams actively combat management techniques that attempt to discipline the methods by which games are developed. As discussed for overtime abuses in the DEMO_MODE for this world, this resistance to process changes is due in part to the origins of many US game developers who get their start while in college, working out of dorm rooms, garages, or basements. Many studios have worked to integrate new methods and practices to innovate in the area of process; yet the lack of experience and discussion about these processes fails to sufficiently prepare developers for the melee that frequently ensues when these highly coupled, complex systems interact.

While "process" as generically conceptualized is where many developers believe the "solution" to work/play issues lies, "scrum" has become the most common. And it works, to a certain extent, except that the interface with external demands made by publishers and console manufacturers, secrecy about game development and the fetishization of interactivity all prevent process from making fundamental change to the industry. The producer of the game *Crackdown* notes how the constant movement from platform to platform created significant difficulties during the development process.

> Over the course of its four-year development, *Crackdown* moved from PC (where it was prototyped), to Xbox (where it was initially intended to stay), back to PC (in preparation for move to Xbox 360), to Xenon Alpha, then Xenon Beta, and at last to Xenon/360 Final. Even on the final hardware, we continued to take hits from significant system software updates every few months. When at last the platform stabilized during the last year of development (post hardware launch), development efficiency increased massively. (Wilson 2007, 29)

This reminds us that, while developers are at the whims of console manufacturers, their ability to control their own process is severely limited by the industry habit of creating efficiencies and standards for each project as it develops rather than standardizing tools and systems.

While it is true that process can improve difficult situations, it cannot account for the myriad influences that shape the socio-technical milieu of game developers. As with pipeline and engineers' tools, Scrum requires localization at the level of each studio. It is a conceptual framework that must be worked out by each studio and must frequently be modified for each project. Scrum is a process and as such, doesn't begin to address the necessary technological issues associated with the construction of pipelines and tools that accommodate those processes. The pipeline is to data, assets, and code as Scrum is to the collaborative, organizational, and communicative practices of developers. The continued norms of secrecy demand that game developers must experimentally figure things out on their own, which makes the broader success of all projects difficult for game developers. While Scrum offers promise, it (or any technique) will remain inherently limited without mechanisms for institutional learning.

World 5-2: The Importance of Passion

Talk to game developers for even a short time, and you will quickly hear that the game industry is "more social" than other industries, and that networking is one of the most important aspects of "breaking in." Gaining access to the industry has become a game in and of itself. Nearly every game developer has a story to tell about how they managed to break into the industry and is more than happy to share it. I often joked with new developers that "you've got to make games, before you can make games," which usually elicits a laugh, or a story about the demo reel, engine, or portfolio that they were certain had something to do with their hiring.

Box 5.2

#: SET DEMO_MODE 1
AUTHOR_DEV_DILEMMA: Many developers talk about how it isn't a requirement that people crunch or dive so deeply into their work, rather it is just what happens. That there is so much passion, drive, and a lack of other interests early in a developer's career that it is easy to leap into the work with little regard for the rest of the world. In other cases, developers do identify that there are times and situations where crunch was required, for one reason or another push forward and work for extended periods of time.
Casey: So what causes that push to produce so much?
TECH_ART_1: I think it's a lot of things. Some places, it may very well just be the love of it. So, take *Bungee* for example. They're making *Halo 2* right now, and they know they're going to make millions and people are going to buy it and its going to be awesome. So I can imagine many employees just don't mind being a part of that. But, I've seen places where working tons of overtime was purely to make up for management miscalculations or bad scheduling or demos for higher-ups that are sprung on people. Bosses could have just said, "No," to things like that, but they didn't, because it isn't them working really hard. In other cases I think it is just immaturity with regard to running businesses.
Casey: Do you think that is changing?
TECH_ART_1: It's also because it is such a competitive industry to get into. You feel like if you don't go above and beyond all the time that you're going to get weeded out. But I think when you have management saying, "You need to be here for twelve hours a day, just for the sake of it," that you suddenly have a problem. Some people who are driven are also efficient and they don't need to do that. But you can't just put it on the managers. There is a culture that supports it. Not to mention the kind of screwing around that goes on. I mean, at one place I was at, we would play *Battlefield* all day. Ugh. Bad decisions.
AUTHOR_DEV_DILEMMA: A competitive industry, dependent on talented and committed individuals, capable of providing them with space to do interesting work is enough for many to slip into life and work styles that actually prevent long term engagement with the craft of game development. Thus, in some cases abusive work patterns derive from passion and in others from an employer requirement (and those reasons may very well differ from person to person on a given team). Yet, it all falls into the same category: overtime. There exists a culture of overtime that is simultaneously requirement, expectation, and simply a product of passion. "If you aren't passionate enough to make games, then maybe you shouldn't" becomes a consistent mantra. Certainly, there are people willing to take your place if you can't or won't work long hours.
#: SET DEMO_MODE 0

Nearly every US game developer who has broken into the game industry has done so by making games, or something as close as they can get, on their own time. While this means that game development companies benefit from new hires being moderately more knowledgeable of what it takes to create videogames, it also means that the industry imports some of the work habits of college students into their studios. Procrastination and an "it will come together in the final hours," mentality frequently prevails. Some developers even bring college sleeping and eating habits into game companies. Late to rise and late to leave can frequently become the *modus operandi* of a studio if young developers are left to their own devices. For many, game development is the interlude to a host of other work- or school-related activities. It is quite literally play for those looking to break into the industry.

I myself broke in during the transitional period, where game programs at schools were just starting to come into existence. Maybe this is why I still tell high school students who want to make games that they should:

- get a degree in CS, or art, or business, or whatever at a reputable four-year school
- work on lots of games, preferably with a game development club, in their spare time while in college (start a club if it doesn't exist)
- and network the hell out of every conference or IGDA meeting they can get to

I guess I'm old-fashioned, but I still think it's the best way. Because, believe it or not, you may change your mind about making games while you're in college. And it would be nice to just say, "Okay, I'll just stop making games in my spare time and going to conferences," as opposed to having to change your major, maybe change schools, and worry about all those extra credits you took that are not going to help with your new degree. (Kazemi 2007)

In addition to importing poor time management techniques from college and the sense of play that surrounds the production of games, this perpetuates the pervasive culture of secrecy that dominates the industry. Students are expected to "make games" without any knowledge of how they are produced in the actual game industry. Breaking in perpetuates the culture of secrecy. Since these students' learning process is divorced from insight by those with experience working in the game industry (either because they are unwilling or unable to share information gathered through experience) many developers view their own solutions to common development problems as the only possible ones. Because US developers have not had an opportunity to make games in the context of work, the ability to discipline or improve their practices is rarely taken. When game development is only a hobby, there is no drive to discipline the practice or understand why paying attention to practice is important. Young developers who find themselves at studios that invest time and energy into developing best practices, tools, or production methods frequently rail against their newfound restrictions and parameters. These developers do not last long at established studios and instead strike out on their own, often making the same mistakes as they muddle their way through the process of making games. Learning how to make large-scale games, or games that can be easily modified to meet the demands of numerous actors, is not done until one reaches the workplace.

It is in the transition from play to work that developers begin to take seriously the importance of practice. Constructing pipelines and tools are rarely (if ever) documented or discussed by developers. Even where those conversations are occurring, most young developers fail to even know they should be paying attention. This is not unlike the work of engineers or technicians more broadly demonstrated by ethnographic fieldwork of these organizations (Orr 1996). Each studio becomes responsible for developing sets of practices, ones usually based on those who started the studio, and their collective experiences.

In contrast, many Indian game developers go straight from their undergraduate education to game development companies often without a portfolio or game development experience. While these developers must learn the ropes of game development on the job, they do so in a context of work. This frequently makes it easier for them to understand game development as work, rather than only as a personal passion. In the Indian market, it is desirable for developers to have experience making games, but most companies assume that a significant amount of training will be necessary; as a result, developers in emerging game industries don't experience breaking in as US developers do. Even Western European developers seem to feel that breaking in is an experience of American developers.

The disadvantage of a less rigid barrier—of an industry in which most new developers are completely inexperienced in game creation combined with game developer's secrecy regarding the everyday work practices that speed and enable development—is that many young game development companies are on uneven footing when it comes to development operations. This tension is greatest when it comes to the full game development process; the development of a game from start to finish, often referred to as the "completion of a title."

The "titles" (games) in which game developers have appeared in the credits tends to be the form of cultural capital that gets them hired when changing studios. For developers in emerging industries, like India, this can be problematic since the same networks that govern studios also govern individuals by default. Studio heads may have willingly traded away the ability to publicize their involvement in the production of a game for greater monetary compensation, which means the studio and its developers get no credit for the title. While many of my Indian informants joke that the Indian legal system makes it difficult to enforce some of the limitations rendered by NDAs (including not insisting on credit for work done in international studios), most of my informants would not consider the possibility of breaking the agreement. This issue of involuntarily producing uncredited work is of crucial concern for game developers, which has made it one of the many issues in which the IGDA invests time. Game industry veterans Feil and Weinstein introduce the association's recent efforts to standardize crediting practices.

> Crediting in the game industry has become a hot topic in recent years. As development teams grow bigger and outsourcing becomes more prevalent, the informal crediting procedures used become increasingly insufficient to describe each developer's exact role within the development process. Additionally, the non-standard naming procedures for job titles that have thus far characterized the free spirit of the gaming industry have now become a liability for those who wish to prove their skills when moving from one company to another. A movement to standardize crediting procedures and titles has never been more needed.
>
> The IGDA Credit Standards Committee is a group of volunteers who have come together to study, document and propose voluntary game industry crediting practices that properly recognize those responsible for the creation of games. To do this, we are creating two documents: one which details the current methodology of credit assignment as well as catalogs a set of the most accepted job titles found within the industry; and another report to propose a set of "generally accepted practices" which can be adopted voluntarily by developers within the industry to resolve difficult crediting dilemmas. (Feil and Weinstein 2006, 3)

Having to depend on credits or titles as the measure of a developer's worth, especially given the controversy and lack of standards around the practice, seems a weak metric for measuring the vast amount of work that goes into the production of a game. This ultimately suggests the question, "What, precisely do studios desire of those working within these secret spaces?" The answer, perhaps, lies in how those secret spaces are defined, and who is allowed to play within them.

This text has touched often on the idea of work/play, a concept that speaks to the kinds of activities that occur during the work of game development. One reason that we link these two experiential and professional concepts, other than the fact that developers play games and do work to make games, is that work/play provides a mechanism that brings other kinds of effects back into the conversation. Work is laborious, and game development certainly is no exception, but it is also enjoyable in ways that speak to play. The imagination or imaginative capacities of play have a great deal to do with our linking of work/play. As play theorists note, "imagination, flexibility, and creativity" of the "play worlds" link to narratives of innovation and progress (Sutton-Smith 1998, 11); yet the ability to be creative, flexible, and imaginative on the job is not all that new. Nor are the problems of both parsing and harnessing work/play. Many in game development and new economy work more generally have attempted to create a hierarchy of work, one that separates the imaginative capacities of one set of laborers over another. While it is true that there is an important difference between creativity and imagination for personal amusement versus creativity and imagination as aspects of work activity, the linking of work and play as something particularly unique to new economy, or game development work in particular, seems premature. Rather, it has likely long been an aspect of creative collaborative practice pre-dating the new economy.

One part of what turns play into work, especially in game development, is skill. The necessary "skill" of workers in game development is difficult to place, seeming to lie somewhere between intuition, technique, experience, technical prowess, and artistry.[6] Game development is varied in its capacity, a fact that is frequently ignored in favor of focusing on the engineering tasks of game development. The work of creating games is highly complex, but also it can be highly repetitive. I was continually impressed by the technical feats my informants could accomplish, and simultaneously surprised by how much repetitive work they tolerated. The concept of work/play allows the cultural analyst to conveniently ignore the complexity in which these things coexist without attempting to understand their broader context. Work/play, when considered as a muddling practice of stops and starts, demands interest and drive (or perhaps passion) but also requires "skill" to inform those activities.

Of course, the concept of "skill" is problematic, as it serves as another means by which access is managed. Rather than using more concrete metrics, studios privilege this amorphous and immeasurable quality in developers, which can act as a convenient means for eliminating or preventing the acceptance of alternative perspectives. The skills necessary for success in game development are wrapped up in a culture of secrecy. To find out how to make games, developers must make games. This circular logic elicits mirth from developers, precisely because of its contradictions. To prove yourself capable of working in the game industry, you must work in the game industry. Even interviews can exhibit the strangeness of these requirements. Riddles, trick questions, and other abstract tests can be used to determine one's "fit" in an organization. Most important, at least in linking our work/play system to those systems of playful relation that actually drive it, is the closure and secrecy of social networks in the game industry. The presumption of, "This is for us, not for the 'others;' what the 'others' do 'outside' is of no concern of ours at the moment," makes it convenient for developers to ignore the insights of others, occasionally cherry-picking ideas from other worlds without really examining the consequences of those ideas. An insular industry serves as a kind of cultural blinder for developers, blocking the industry from outside view (and insight) and trapping developers within its cultural and technological walls. A problematic result of this blind isolation is that developers distance themselves from alternative perspectives: "Inside the game, the laws and customs of ordinary life no longer count. We are different and do things differently" (Huizinga 1971, 12). This sense of being apart from, as mentioned in World 2, allows developers to assume their worlds as uniquely distinct in the world of software-mediated creative work. This evokes Huizinga's observation that play often attempts to wrap itself in a sense of mystery and assertion that the real difference lies simply in the demand of play to call itself different. "The exceptional and special position of play is most tellingly illustrated by the fact that it loves to surround itself with an air of secrecy," is of critical importance to our understanding of work/play (Huizinga 1971, 12). We do things differently here; do not expect your ways to be our ways.

It is possible to situate the culture of secrecy of the game industry in a sense of play, but given its temporal persistence, it seems inadequate. In more than one instance, informants referred to the craft of game development as a kind of technological "dark art." The idea that the craft of game development is almost a kind of witchcraft or sorcery places it more centrally in the culture of the community. There are numerous anthropological examples of how witchcraft and sorcery are central to "control[ling] flows of power in a society" and often "incorporate general concerns about uncertainty, scarcity, and risk" in relation to ambiguous power flows. (James 2012, 50–51). The close control of social relations and the desire for creating order in light of complex forces that destabilize the community makes intuitive sense. In many respects, "dark sorcery" and its craft are an important means by which cultures are "capable of defending [themselves] against the depredations of the outside world" (Whitehead and Wright 2004, 7). By rendering aspects of work and social relations opaque, those operating on the inside protect themselves from outside influence. Yet, the maintenance of the culture of secrecy among game developers is not so actively perpetuated. It more accurately fits into the model of a "public secret," or "that which is generally known, but cannot be articulated" (Taussig 1999, 5). It is with this in mind, that I "point out the obvious" of mythologies surrounding game development, for even the "obvious needs stating in order to be obvious" in this case (Taussig 1999, 6).

World 5-3: The Game Develop(er/ment) Mythology

These cyclical arguments of needing to open up, yet rejecting such an act of subversion in the culture of play, hearken back to the "breaking into the industry" narratives of informants, and the entire sub-genre of game development community writings about how one can gain entry to the restrictive networks of access. While it is true that work/play is imaginative, interesting, and desirable, it seems that the way it wraps itself in secrecy and closes off networks of access further elevates its status and desirability. The levels of secrecy and networks of access pervade numerous aspects of the work/play of the videogame industry. The mythology that surrounds the game industry about what game development is, who it is for, how it takes place, and its broader political economic context all enter into this equation.

The secrets of manufacturers, even when revealed to a licensed engineering team, are protected by restrictions on what may be freely shared outside of an individual studio. Chances are, even if two studios are covered by the same contracts and non-disclosure agreements, they will be reluctant to speak in specificities for fear of transferring some specific piece of "proprietary" information. The threat is not only of revealing secrets that might give a rival studio an advantage, but of angering the publisher or manufacturer who controls future contracts. Artists who move from one company to another will encounter entirely new, though perhaps quite similar, technical systems that enable their work/play. The immense categories of work covered by copyright, non-disclosure agreement, or corporate contract encourages the fallback position that everything is secret. Names of characters, basic game mechanics, story lines, tools, model requirements, engineering standards, processes, pipelines, organizational structure, clients, publishers, and hardware being worked with all begin to fall into the category of "secret." And because ongoing pervasive secrecy within and among each studio compounds the historical legacy of secrecy in a walled off industry, potential answers about how or why systems have been constructed in the way they were remain unknown.

The importance of the secret, combined with an average career lifespan for game developers that hovers somewhere around five years (Staff 2007), results in a situation that resembles a perpetual startup company machine. Experience is crucial to the game developer's work, but it is the very thing that has proved elusive for the game industry. Secrecy and inexperience leads to continual reinvention of the industry in terms of methods and technologies. Those companies that do succeed at creating games are often purchased and incorporated into the increasingly secretive inner circles of the videogame monoliths and the acquired knowledge locked into powerful, legally circumscribed spheres. Game development has no theoretical foundation that can be simply applied; and even if an all-encompassing theory of game design did exist, all theories are "models or tools," and it is the technologist's ability to "apply theory through recognizing situations as similar" (Turnbull 2000, 43) that would allow the universal use of that knowledge. Few developers have access to shared information, even fewer have lengthy careers, and therefore the opportunity for industry learning, maturity, and advancement is lost.

Most conversations with developers tend to focus on "sustainability" or the ability for the industry to exit its state of perpetual startup. This "sustainability question," which is what drives much of developers' anxiety about the industry, has very little to do with monetary stability. The massive amounts of money swirling around the videogame industry at the moment, though staggering and likely unsustainable, are not the elements that concern developers. Their unease focuses on the long-term viability of making games as a professional craft.

In part, the sustainability question is linked to a widespread lack of understanding of the myths and realities of the videogame industry. The complicated matter is that, in many respects, the industry, as it has currently constructed itself, depends on widespread belief of the myths noted in table 5.1, which encourage or hinder new employees and new investment.

Table 5.1

Common myths about the game industry

Myths
1. You get to play games all day.[a]
2. You get to make the games you want.
3. You get infinite time and resources to make a game.
4. Every game makes millions of dollars, and so do game developers.
5. Games are for kids.

[a]And, by corollary to this myth, should self-ascribe to being "gamers," which has significant implications for the diversity of individuals seeking to make games.

These myths, as toxic as they are to the industry's ability to attract and retain talent and capital, persist in part because of the industry's penchant for secrecy and in part because developers simply don't do enough to discourage these myths. For the most part, the all-day game-playing myth has begun to wane in US conceptions of game development, though popular culture does still fall back on this myth, as demonstrated in a movie titled *Grandma's Boy*. The story line's basic premise is a weed smoking, unmotivated yet brilliant Q/A developer who plays games all day with his coworkers. In his free time (of course) he is able to develop a game for the Microsoft Xbox completely on his own and without a DevKit, SDK, debugger, computer, or actual resources of any kind. His "development" process also appears to be his playing a videogame that his grandmother also plays on several occasions, offering her feedback and adjustment of. Even game development programs at schools and universities will fall back on this myth of constant game play in an effort to recruit new students.[7] In the commercial spot, two twenty-something men gaze toward a screen invisible to the viewer, controllers in hand. "Oh, better hurry up, the boss lady is coming." The boss stands in the doorway and asks, "Have you guys finished testing that game yet? I've got another one I need designed." To which one replies, "We just finished level three and need to tighten up the graphics a little bit." As the ad closes out, the two men exchange glances and one confesses to the other, "I can't believe we got jobs doing this." The other replies, "I know, and my mom said I would never get anywhere with these games." People who talk or write about videogames, however, are beginning to understand at a basic level that the actual work of making games involves tools that engineers, artists, and managers have already used for years, with new custom tools filling in the holes when necessary.

Box 5.3

#: SET DEMO_MODE 1
AUTHOR_DEV_DILEMMA: Much of game development isn't actually the creation of new and exciting titles, but an attempt to capitalize on the movie industry model of sequels upon sequels of any one given intellectual property, except that game studios have significantly more work required between sequels.
Casey: I know a lot of game developers worry about sustainability.
STUDIO_CREATIVE_HEAD: [T]here is no good way for publishers to leverage their content beyond a certain point. So the game gets

(Continued)

```
released. It may go to greatest hits, which is actually a pretty
good model that keeps games on the shelves longer, but then it is
gone. And often it is because the technology has moved on and the
platform you played it on is outdated. Games have no model of:
theatrical release, pay-per-view, premium cable, DVD, broadcast
TV, cheap DVDs in the bin at Walmart. There isn't a model like
that. Nintendo does a brilliant job of bringing out a new hand-
held and putting Mario back out on it and selling a few more mil-
lion copies. They are one of the few that have figured that out.
But, that isn't an industry model. Leveraging your back catalogue
of IP is crucial for long-term sustainability.
AUTHOR_DEV_DILEMMA: Because of the technologies, tools, and sys-
tems involved, development companies must toil to recreate titles
with entirely new technologies or for different hardware plat-
forms, which is often a much more difficult task than is acknowl-
edged. Many game development companies work on other company's
intellectual properties other than their own.
#: SET DEMO_MODE 0
```

Cloaking the day-to-day reality of long hours and tedious work on sequels, the myth about making whatever games you want is perpetuated far and wide, even among some current game developers. Only those who have worked in the game industry for years and developed a jaded/realistic attitude toward their projects know that they rarely, if ever, make the games they want. The majority of the time, the studios work on games envisioned by others. The developers' assignment is to make someone else's vision a reality (and as much fun for someone else as possible). The creative license myth, however, is critically linked to the myth that most games make millions and easily recuperate the associated costs. Frequently, the most successful games are breakout hits, which are different and new, or those that capitalize on an already established market and franchise. Most of those established games are developed on contract with third party development companies.

Just as developers don't play games all day, and rarely work on games of their own choosing, they do not have unlimited time or resources. While, from an external perspective this may seem true of certain companies such as Bungie, Blizzard, or Id (all of which are legends in the game development community), the reality is that even these companies who have the power/license/cachet/ability to respond to questions about schedules with "It will be done when it is done" are internally quite aware of their time and budget limitations.

And those budget limitations do not include huge salaries for developers. Though no game developers I spoke to ever imagined making millions of dollars based on their work in the videogame industry, the myth of the wealthy game developer remains: "The computer industry's garage-to-riches myth fuels the hope of instant success despite evidence to the contrary. . . . Indeed, younger workers are well on the way to believing that taking entrepreneurial risks is necessary to building careers. This is the legacy of 1980s-era enterprise culture and corporate restructuring" (Rogers and Larsen 1984, 154). The perception of development as wildly lucrative hurts the industry because it ignores the numerous factors that may lead to a startup studio's demise, simultaneously glorifying risk and withholding a critical examination of who often then receives the majority of the payoff.

With so many game companies closing or having layoffs, a perception exists that, regardless of how well the game industry is doing, getting another job will be difficult and as such, there is a "disincentive to exit during difficult economic times" (Neff, Wissinger, and Zukin. 2005, xx). Further, many game developers exhibit an entrenched commitment to notions of "meritocracy," within the game industry. If game developers are any good, they can go anywhere. If developers are without work, then they must not be very good. These theories enable "continued attacks on unionized work" (Neff, Wissinger, and Zukin 2005, 317–330). The two aspects of employment and job search combine to obfuscate systemic issues in the game industry, transitioning blame from external aspects to individual ones. For example, there is an assumption that a developer's performance and compensation has a direct relationship to an individual's skills. If a developer is unsuccessful or not paid well enough, the problem is a personal one; there is no possibility that structural issues might be condemning developers. Put in the language of the videogame industry, if you crunch it is because you did something wrong. And if you object to crunching, you shouldn't be in the game industry.

The final myth, concerning the idea of games as child play, is in many respects the most complicated, especially given the contrast between India and the United States. As mentioned in World 2, for many Indian developers, working in the game industry is not considered a reasonable career choice. US developers, many of whom do not take their family's or parents' advice when considering a career path after college, recognize that their family might lack understanding and support for any career, not just for their choice to pursue a game development job. For Indian developers, it is assumed that your family is part of your consideration process in pursuing a career. Many game developers take their game job against the will of their parents, and eventually translate their experience into positions at companies like Google, Microsoft, or Infosys. These more "reputable" companies alleviate the familial strife (and benefit from the game industry's poor reputation in Indian culture). The perception that games are childish significantly impacts the ability for students to move into these positions. Given the choice, when Indian workers are willing to take the risk of offending their families to develop games, many pursue work with established companies over local burgeoning game studios. The threat of offending one's parents is risk enough, no need to bring in additional risks associated with small companies and creative work. The ability of these new studios to push themselves onto the global scene is hindered by the same perception of illegitimacy, or lack of rigor of videogames. Given the fact that many videogame publishing companies are establishing development studios overseas, there may be a decline in the conception that videogame development career paths are not practical. However, such international development will also likely negatively impact the viability of locally started game development studios that will have to compete for employees with more well-known names, such as Sony Computer Entertainment, Microsoft, Electronic Arts, and Activision.

All five of these myths continue to hobble the professionalization of game development practice. Yet, developers are not working to fight or even discourage unrealistic views and expectations. In fact, in some cases developers even promote the myths. This emerging system of relations between corporations and game development studios plugs directly into a system of labor relations found in many "cool" industries. Sociologists studying new media companies frequently find that knocking workers down builds profits.

> The labor relations within cultural production provide global capital with a model for destabilizing work and denigrating workers' quality of life. The cultural workers in fashion modeling and new media work long hours, networking even while they are schmoozing and boozing, constantly try to improve their skills, and live with a high degree of insecurity about their income and employment. These workers now directly bear entrepreneurial risks previously mediated by the firm, such as business cycle fluctuations and market failures. Popularized in media images of cool jobs and internalized in subjective perceptions, this work creates a model of labor discipline for other industries to follow. Moreover, given the ethnic and gender characteristics that have been associated with entrepreneurial culture, the effect of these changes will exacerbate persistent social inequalities. (Neff, Wissinger, and Zukin 2005, 330)

As workers are more frequently asked to bear the consequences of a denigrated and destabilized quality of life, they are asked to bear greater amounts of risk once born by the organization. Yet, because there is a perception that the individual can mitigate this risk through personal passion or perseverance, it becomes permissible to blame the individual for any feeling of insecurity. This plugs directly into a kind of libertarian worldview that dominates many of these industries. While workers are capable of working within this regime, it works to their advantage, until it is no longer possible, at which point it becomes their own fault and they can easily be blamed for their falling behind.

The myths that perpetuate an idea of what game development is prevent it from being viewed more broadly as worthy of consideration as a profession or realm of expertise. It encourages a mass of interest in the field that may prove unable to actually provide the requisite necessary areas of expertise. Thus, workers find their everyday working worlds destabilized because they are perceived to be without expertise or profession. The "coolness" of the work makes it "easy." Anyone can be cool, right? But, as I've noted throughout this text, game development is a highly technical, creative, exploratory and fraught process that is far from easy. By destabilizing workers and their labor in pejorative ways, the risk is further offloaded on the individual from the broader context of the game industry.

World 5-4: Designing the Perpetual Startup System

As the myths surrounding game development work bind developers and limit the industry, the risk associated with developing new ideas has been offloaded from companies onto workers. The image of game developers as pushing the envelope, realizing their own creative ideas, having fun, and getting rich has led them to willingly trade a sustainable industry for the negligible possibility of making it really big. Corporations capitalize on what can only be termed "adventure capital" of game development workers by cherry-picking the best companies for consolidation and acquisition in order to benefit from the risks they've already taken. Risk taking is handed off as a badge of honor for developers, and the prize is collected by the companies who hire them. VV's acquisition by Activision is one such example. Throughout my fieldwork at VV, Activision acquired eight different studios, four of which were later closed.

The primary consequence of this structure is what I have called the "perpetual startup cycle" of the videogame industry,[8] a situation where most of the risk associated with expanding markets and developing new IP is borne by small startup companies, those frequently with the most to lose. Large companies with enough capital to afford risks frequently eschew them in favor of ensuring good quarterly reports, as can be seen in publishers' and manufacturers' overwhelming preference for derivative games and licensed IP. When a startup is able to prove itself capable of producing value, it is acquired, so a large, established company can milk the value out of the risk-taking resources. Further discussion of this phenomenon can be found in World 6. Those who tire of this cycle of risk being relegated to small development studios while reward comes to large corporations often perpetuate the loop themselves by leaving and starting new companies, taking risks, and again pushing the industry in new directions. Again the cost associated with trying something new is borne by those least able to. Startup companies, with their demanding and tiring work environments (which our earlier DEMO_MODE showed can be particularly daunting for anyone with a significant other at home or a social life or hobbies outside of the office), drain the adventure capital of their contributors in order to sell it to large corporations.

In many respects adventure capital resembles and is infused with the same "venture labor" (Neff 2012) that can be found in fields similar to the game industry. What I think distinguishes adventure capital is the sense of play and almost flippancy about risk that imbues many game developers. In games and in particular digital games, players fail repeatedly. Failure is not just part of the game, in many cases it is the name of the game. You fail and learn. Fail and learn. The "adventure" aspect of this is of course also a tip of the hat to the 1970s adventure style text-based games that borrowed, built and learned from one another in the early game industry. Thus, adventure capital is part venture labor based on intense risk and accidental entrepreneurship. It is also more playful and assumes a kind of serial failure that is in some ways different from the kind of serial venture labor that one encounters in the new economy.

If you watch the videogame industry (or any industry, for that matter) in this first part of the twenty-first century, you are bound to notice that there are many acquisitions occurring. New media industries including the videogame industry exemplify of this kind of activity. "'In this industry, because it's changing so fast, you're lucky if you're in the same job for a year,' says a producer for a corporate, online retailer" (Neff 2005, 326). Even the once famously independent game studio "Blizzard Entertainment, Inc." was acquired by Vivendi/Universal, itself a consolidation of two large media organizations, and later sold off to Activision.

My primary field site in the US transitioned from independent/third-party developer to in-house developer in the winter of 2005. While Castells has written about the "crisis of the large corporation" (Castells 1998, 167) and the "resilience of small and medium firms as agents of innovation and sources of job creation" (Castells 1998, 167), the fact remains that "small businesses are less technologically advanced, and less able to innovate technologically in process and in product than larger firms" (Castells 1998, 167). The large publishing companies that employ the vast majority of experienced game developers in one way or another (either through direct ownership of studios or through licensing deals) end up doing less innovative work as judged by the comments of my informants, despite larger companies' higher levels of technological and experiential capacity. Inter-firm linkages are "the multidirectional network model enacted by small and medium businesses and the licensing-subcontracting model of production under an umbrella corporation" (Castells 1998, 172), though they are intra-firm in most cases because they are the firms already capable of doing much of the work. As game companies gain development experience, they often forget the adventure capital already invested in the infrastructure that enables development. As such, established game companies are quite different from their younger and less experienced counterparts. Anthropologists of scientific and technology production note the increasingly complex interconnections between organizations: "Production increasingly takes place within larger organizations, each of which is more likely to include multiple locations, many of which in turn are in different regional, national, and cultural locations. Moreover, more permeable organizational boundaries mean production occurs within technical and social networks which cross company cultures" (Hakken 2000b, 770).

This structure, especially in that it involves multiple corporate cultures, doubly complicates the situation for game development companies in countries other than those with already established networks. As has been demonstrated, game development work, with its numerous disciplinary fault lines and technological complexities, is procedurally and structurally complex. Without established peer networks willing to share people, information, or experiences, new companies have a predilection for collapse. While the acquisition behavior in India is in full swing, networks of access limit it. Take for example, Indiagames, which UTV Software acquired. Indiagames had inserted themselves into the networks of mobile phone networks, and had developed connections that UTV Software could use to its benefit (hence the acquisition). RedOctane—India (part of the US company

RedOctane, which published Harmonix's wildly popular game, *Guitar Hero*), on the other hand had no ability to develop games for console systems until it was acquired by Activision. After the acquisition they were able to use the networks within Activision to begin working with Nintendo's DS systems. Hakken's observations, then, that production occurs within permeable and flexible networks is correct, but also limited by the connections of those networks. Permeability does not equate with accessibility.

Of course this does not begin to address the complexity of dealing with publishing companies that in many cases have no knowledge of an emerging industry's internal market. These companies only know that they are supposed to be moving their "activities of production, consumption, and circulation, as well as their components (capital, labor, raw materials, management, information, technology, markets)" (Castells 1998, 77) toward being "organized on a global scale, either directly or through a network of linkages between economic agents" (Castells 1998, 77). Companies' options include developing games internally and self-publishing (which limits them to distributing online for the personal computer, Web, or more recently on mobile "app" stores), developing games with contractual obligations with other companies, or being acquired and developing games as dictated by the parent company. If publishing companies have become "conservative" in the United States, they are doubly so in emerging markets, which they often understand as only capable of consuming First Person Shooters, Drivers, and Massively Multiplayer Online games.

Franchises derived from local markets (Bollywood or Hindu legends, for example) that could potentially be lucrative are too risky for US, European, or Japanese producers whose limited understanding and low risk tolerance demand that local development and productions remain small and independent. Publishers and manufacturers who do dip their toe into such markets ship only existing titles and refuse to authorize the development of titles specifically for those markets made by people who know the market; others will simply refuse to market their games or consoles in the region. This avoidance of taking risks to extend into new markets is an example of how the industry is thwarted by the model in which small studios take all the risk and large corporations capitalize only on what's popular.

Like publishers, most manufacturers are hesitant to risk spending much time or energy marketing to emerging markets. Microsoft is a particular outlier in India, marketing the Xbox 360 throughout the country and even partnering with local Indian banks to provide financing to encourage broader adoption among the affluent middle class. Despite this relatively daring move on the part of Microsoft, no authorized licenses for developing games on the 360 have been given to developers in India. Microsoft's games continue to be developed at US and Western European studios. Sony, on the other hand, maintains retails stores in India that have begun selling the PSP and PS3. They have not aggressively marketed their consoles in other regions, and the particularly expensive character of the PS3 makes it a difficult sell in emerging economies. Nintendo seems to refuse to acknowledge India; even in a market with massive support of mobile devices most gamers have no knowledge of the Nintendo DS system, which caters to the mobile market. The only Nintendo systems in India are those that have been imported or brought back by the few gamers who travel between other countries and India.

It might seem that this marketing blackout is just a case of general ignorance of emerging markets. Yet, for game developers, the consequences of these oversights reach far beyond the simple availability of hardware within a market. The importance of being on the right sides of network switches or "part of the club" is dramatically important, though frequently under-examined in the context of emerging networks. Console game manufacturers and publishers carefully monitor those companies that are capable of making games for the game industry. In recent years, the emergence of the iTunes App Store and Android Market has demonstrated the fragility of these assumptions on the part of the game industry. While some people point to these new app stores as "walled gardens," in the game industry, the walls have always been there, and in most cases these new walls are much lower, enabling new companies to emerge as major market players.

> The more countries join the club, the more difficult it is for those outside the liberal economic regime to go their own way. So, in the last resort, locked-in trajectories of integration in the global economy, with its homogeneous rules, amplify the network, and the networking possibilities for its members, while increasing the cost of being outside the network. This self-expanding logic, induced and enacted by governments and international finance and trade institutions, ended up linking the dynamic segments of most countries in the world in an open, global economy. (Castells 1998, 142)

And, at least until recently, in the videogame industry there truly is "only one game in town," a game that rather than being controlled by political elites is working within massive publishing and manufacturing conglomerates. While new distribution platforms have emerged, the primary means of "playing the game," requires access to the more limited networks of the major console manufacturers. These networks have determined the regular rules by which one becomes part of the inner circle; if your game or project falls outside of those rules, you need not apply: "This is because the global economy is now a network of interconnected segments of economies, which play, together, a decisive role in the economy of each country—and of many people. Once such a network is constituted, any node that disconnects itself is simply bypassed, and resources (capital, information, technology, goods, services, skilled labor) continue to flow in the rest of the network" (Castells 1998, 146–147).

The local, the social, and the distinctness of network segments is reemerging rather than disappearing. In game development, an individual is quite close to the thing being produced. While a developer's input into the project may only be a small component of what is experienced by the player, it is often identifiable. This was the very point of resistances on the part of early Atari developers, who were not given credit for their work (Kent 2001). While teams have gotten larger, it is still possible for individuals to contest the flow of the rest of the network. The connections are too many and too diffuse to be completely overridden. Some connection point will remain that can be reconstituted: "In sum, the more the process of economic globalization deepens, the more the interpenetration of networks of production and management expands across borders, and the closer the links become between the conditions of the labor force in different countries, place at different levels of wages and social protection, but decreasingly distinct in terms of skills and technology" (Castells 1998, 254).

Corporate consolidation of videogame studios under umbrella publishing companies has severely limited the ability of developers to gain access to the networks necessary for creating games for consoles or distributing them more broadly. Publishers and manufacturers have both participated in disciplining the particular function of this system, but more in an interest of playing the game to the advantage of their bottom lines rather than out of malice. Many of these publishers began as small game studios railing against the status quo. It would be too simple to say what is occurring now is a deliberate attempt to dismantle game development work; rather, the demands of an economic system dependent upon quarterly results have skewed their perspective. Most employees of publishing companies or manufacturers have a deep commitment to games as forms of art, media, and entertainment. It would be disingenuous to claim that they were all dupes of a system designed to crush innovative game development. We can note more plausibly that the structure of the industry has enabled a continual cycle of externalizing adventure capital, relying on small game development studios to bear the risk and then milking those innovations in the name of profits to meet quarterly results.

World 5 Boss Fight: The Rise and Fall of "Quality of Life"

Why do academics always wind up at the laments of ea_spouse? That infamous post has become work/play pornography, in part because of its accessibility. I suppose academics have not had much opportunity to observe work/play patterns and realities because field site access is so limited. And, a LiveJournal site is so much more readily accessible to the social scientist (Dyer-Witheford and Sharman 2005; Dyer-Witheford and de Peuter 2006; Deuze, Martin, and Allen 2007; Wark 2007), not unlike other forms of Internet porn. Cultural studies analysts have commented on the surface visuals of ea_spouse, and the EA context, but not pushed much further:

> You could be forgiven for thinking this is just a game, but it is somebody's life — as reported in a widely circulated text written by EA Spouse. EA, or Electronic Arts, is a game company best known for its Madden sports games, but which also which owns Maxis, which makes *The Sims*. EA's slogan: Challenge Everything—everything except EA, of course—or the gap between game and gamespace. In the gamespace of contemporary labor, things are not like the measured progression up the ranks of *The Sims*. In *The Sims*, Benjamin could work his way from Game Designer to Information Overlord much the same way as he had worked up the levels below. At Electronic Arts, things are different. Being an Information Overlord like EA's Larry Probst requires an army of Benjamins with nothing to work with but their skills as game designers and nowhere to go than to another firm which may or may not crunch its workers just as hard. As the military entertainment complex consolidates into a handful of big firms, it squeezes out all but a few niche players. Gamespace is here a poor imitation of its own game. (Wark 2007, 044)

This is not to say that the ea_spouse comment has not been an important index, or an important galvanizing point for game industry workers. It most certainly is that. But the fixation on ea_spouse's post draws analytic attention away from the broader issue: why and how does work/play have such a propensity for damage to quality of life? We must recognize that EA and ea_spouse's spouse are positioned in a broad system that encourages practices that enable poor quality of life.

It is easy to fixate on ea_spouse. Erin[9] is a superb writer. However, the fixation on ea_spouse, even in positive ways, draws our attention from the reality of working in the game industry that rarely resembles gamespace and, significantly, changes the reason why developers continue to work in an industry that places such demands on its workers. The desire game, "I want, and will pursue" strikes so clearly at the issues faced by videogame developers. Those people that pursue game development do so, often with a fervor that is easily co-opted into forms of labor that are unsustainable, driving workers from the industry and losing their accumulated experience.

Box 5.4

#: SET DEMO_MODE 1
AUTHOR_DEV_DILEMMA: For some of my informants, sustainability is localized. They look around and wonder what has happened to so many of the game developers that came before them. With an industry now nearly thirty years old, there should be more developers in their fifties and sixties. Yet, the average age of those working in the game industry still hovers in the early thirties (Gourdin 2005).
Casey: So how does that work long term?
TECH_ART_1: That's what I'd really like to know. You really can't look around too much and say, "Where are all the fifty-year-olds making games?" When I'm forty-five or fifty am I going to be valued or that much better because of my experience? Because, someone that was making games in 1982, a lot of what they know means less now. It doesn't really say anything about their relevance now. If that cycle of the industry re-inventing itself every fifteen years or so, where the technology changes completely, then that is a pretty frightening possibility. I mean, I'd like to think that I can keep up and retool, but it takes time and work. It's like when the 8-bit era ended and everyone was a pixel artist and say you didn't have an aptitude for 3D—you were screwed. Not to mention that you have all these seventeen-year-old kids that learn software well, but don't know shit about art. But the industry gets into this big roll-over periodically.
AUTHOR_DEV_DILEMMA: There is an element of reinvention that the game industry does periodically, where old techniques and technologies are traded for newer ones. Yet, this has become less common. Even when new "generations" of consoles are released, there is more similarity to older versions and fewer processes are jettisoned completely from day-to-day practice. But the question remains, "Where are all the game developers?"
#: SET DEMO_MODE 0

Some might assume that all of this crunching is no fun, right? It certainly is true that 50 percent of game developers leave the videogame industry in fewer than ten years (Gourdin 2005). Nearly a third drop out before their fifth year (Bonds et al. 2004, 30–31). Many communication scholars assume that game developers are then just dupes of a post-industrial system that exploits them, which is a conclusion Kline makes convincing.

> All this casts doubts on the myth that game making is 'fun.' Such labour does not live up to rose-coloured post-industrial visions of knowledge work. But nor does it match the straightforward picture of deskilling and degradation painted by the neo-Luddite left. What emerges is more contradictory. The creation of a new creative high-technology industry has required management to recruit a post-Fordist workforce whose control requires the use of techniques that are very different from the rigid routinization and top-down discipline of Fordism. They involve a high degree of soft coercion, cool cooption, and mystified exploitation. (Kline et al. 2005, 201)

But, if the game workforce is coerced, coopted, or exploited to develop games as they do, then why do we find similar work/play characteristics among people working for free? Sure there is soft coercion and cooptation, game studios are corporations. But it doesn't make any sense to argue the developers are being controlled, because despite these conditions, and beyond simply the "cool" factor of it all, people are driven by their jobs. There is something about the intellectual, visual, collaborative aspect of it that hits at something deeper, as anthropological studies of Free/Libre and Open Source Software workers demonstrates. Deep-hack mode does not draw just hackers (Coleman 2001). It affects artists, designers, graduate students, and many others. Work/play has tapped into something that, when allowed to drive to its own beat, does become "mystified" exploitation.

> Just as the military industrial complex once forced the free rhythms of labor into the measured beat of work, so now its successors oblige the free rhythms of play to become equally productive. Alan Liu: "Increasingly, knowledge work has no true recreational outside."* The time and space of the topological world is organized around the maintenance of boredom, nurturing it yet distracting it just enough to prevent its implosion in on itself, from which alone might arise the counter power to the game. (Wark 2007, 172)

This is a different set of desires, the "phenomenology of the zone." I wonder, however, if all of these games are worth playing? What is the payout? What is being pursued and often at such risks? Why then do my, as some would say, "mystically exploited" informants keep doing it?

"So far so good, but what actually is the fun of playing? Why does the baby crow with pleasure? Why does the gambler lose himself in his passion? Why is a huge crowd roused to frenzy by a football match?" This intensity of, and absorption in, play finds no explanation in biological analysis. Yet in this intensity, this absorption, this power of maddening, lies the very essence, the primordial quality of play. (Huizinga 1971, 2–3)

Why are we continually testing ourselves with these games? The question becomes: how can we hack, or "exploit, refigure, and thrive off those social contradictions related to technology, contradictions that emerge more palpably in the tense intersection between liberal values and a neoliberal knowledge economy" (Coleman 2005, 46) to better serve developers work/play needs? Perhaps we can turn, at least in part, to hacker practice as a site for inspiration.

This means a hacker will at times dutifully respect a system of logic while, in other instances, he will blatantly and with succulent pleasure disrespect it, either for the sake of play, exploration, making a political statement, or to accomplish the immediate task at hand. As often as one can find hackers distorting language, laying bare its contingent nature, one can also find hackers who dutifully respect the formal rules of grammar and praise its internal or deep logic with the same incisive precision they treat object variables while programming. Based on elements like individual stylistic preferences, one's ability to manipulate form, and especially the social context of activity, the hacker attitude toward form is relational, oscillating between respectful awe and playful irreverence. (Coleman 2005, 213–214)

Coleman uses hacker practice as a site for her own theoretical innovations, using the term "irony" (in close proximity to historicity) to, "simultaneously accentuate the existence of powerful systems of coercion or hegemonic institutions, as well as the ways in that they are intentionally and accidentally evaded" (Coleman 2005, 34) This use of irony is intricately linked to the practice of simultaneously working dutifully within structures of constraint while also disrespecting them, frequently in humorous ways. This type of play is considered "ironic precisely because it is still possible to tease out the sensible or expected elements within the shell of the unexpected" (Coleman 2005, 34).

But what does this ironic play have to do with breaking out of the recursive infinite-loops of AutoPlay? It can best be described as a process of debugging coercive or hegemonic structures, which requires an attention to detail and awareness that cannot be described in any ways as disengagement. Debugging requires an attention to detail and "disrespect" to a system that enables us to examine the inner workings of a system. Often times debugging involves "stepping into" systems that were previously closed or considered outside of the frame of interest. This is complicated when we are on the "bleeding edge" as "urgency" is compounded by "technological complexity" (Pentland 1997, 118).

This stepping into the system's structure is what separates the hacker from other system (structure) inhabitants. This is how we can escape the mechanism of AutoPlay, rather than being absorbed into the recursive flow of the system. We can step into the operational mechanics and disrespect them. Perhaps this insolence will manifest itself in humorous ways. Perhaps we debuggers will be labeled "black-hat malicious saboteurs" (Coleman 2005, 45), but it is precisely that ability and desire to disrespect the systematic nature of structures that drives us to make changes. This is precisely what it means to transition from being a gamer to a game developer, asking questions of the underlying system.

More than anything, the game developers need to rediscover the importance of sharing and collaborating across corporate divides. They need to reinvigorate their ability and desire to write, talk, and share details of their work that they take for granted. In the nearly thirty years of videogame development, game developers have not managed to share more broadly the reality of their work practice, despite the demands that people entering the game industry already be acquainted with making games. Maybe we should demand, instead, that they know (and actually tell them) how games are actually made.

The demolition of the pervasive individualistic barrier will require developers to give up on the secret society. Game developers can no longer afford to perceive themselves or those around them as independent rock stars. Indeed, many of them are extremely intelligent, hardworking, and deserving of recognition for their creative work. The current system tends to recognize too few. With these changes, more developers can be recognized as integral to the process.

Box 5.5

#: SET DEMO_MODE 1
AUTHOR_DEV_DILEMMA: In some respects, the yearly Game Developers Conference (GDC) inhabits a strange space in the game industry. It attempts to bring developers together to discuss their practice in ways that share information. Yet, simultaneously, most talks are vetted through legal and layers of management that may completely disable the utility of a talk. In other cases developers throw caution to the wind and discuss those things that maybe

(Continued)

even they were instructed not to, taking on the adventure capital of sharing information that they feel the game industry so desperately needs at their own peril.
Casey: Does that prevent developers from being able to have frank, useful conversations?
DESIGN_LEAD_1: Sure. It is always interesting at GDC, the kinds of discussions you have. They're often at a purely theoretical level. Or, they are just trying to sell you something. "Look at our awesome game! We're awesome!" Or, "Look what we're about to do! We're awesome!" I hate that even more, because they never actually get into any detail. The discussions are purely at this high level. Then there are the folks who rightly claim, or cynically claim, that anyone who's actually figured out anything will never share it with you anyway. Unless of course it's too late, because you'll never catch up anyway. Yeah, that kinda sucks.
Casey: So how does it get better?
DESIGN_LEAD_1: Generally there are people who you have closer contact with. You have a professional rapport with them. You know they're not going to stick it out in public or blog about it.
Casey: So how about information sharing even between studios with the same publisher?
DESIGN_LEAD_1: Kinda sucks. You know, our publisher really touts their independent studio model for success, though I've never had that adequately explained to me why it works. So we have contacts, and I've met folks from other studios. Given the distance and often just convenience, it's just not that easy. We try to rectify that, but there are studios that we work more with than others.
AUTHOR_DEV_DILEMMA: Thus, it is primarily within the established networks of the game industry that information is shared. Yet, even then, the constant time restraints placed on teams means that they are rarely afforded the time to explore these collaborative opportunities.
#: SET DEMO_MODE 0

Fixing the ways in which studios work will also necessitate change in how game development companies recruit new talent. Rather than demanding that aspiring youth "break into" the game industry, there must be mechanisms by which they can be encouraged into the industry. "Talent" and innovation must be fostered rather than demanded from the outset. If developers allow themselves to maintain the existing structure where only those who have already "figured it out" are authorized, they will continue to get more of the same: inexperienced developers who reinvent the wheel, grow frustrated, and leave. Instead, developers must become more accessible. They must begin to share more publicly, and more with one another. They must begin to think of themselves collectively rather than as individuals or individual studios. A sense of "the profession" and culture of the game industry must become something that people actively engage with and consider.

Glimmers of hope exist. A handful of studios have embraced the idea of sharing tools and practices with the community more generally. Insomniac Studios, for example, has developed their "Nocturnal Initiative," which has made publicly available more information about their tools and pipelines. More and more game developers blog and tweet publicly about engineering, design, and artistic workflows. New tools, like Unity, have emerged that place the issue of the "pipeline" at their center, rather than the periphery. Scrum and Agile development methods are being more widely used. More conferences have emerged, particularly among "indie" game developers, where ideas, early prototypes, and methods are discussed in detail. However, even among those that share information, similar constraints emerge and problems emerge. Thus, it seems that secrecy and fears around sharing are a systemic issue. In other words, there is an underlying system (a game mechanic, perhaps) that encourages the system to break in consistent ways that continue to steal adventure capital from those with the fewest resources: rank and file developers.

Publishing, Manufacturing, and (Digital) Distribution

World 6: Actor-Networks of (In)access

Box 6.1

#: SET DEMO_MODE 1

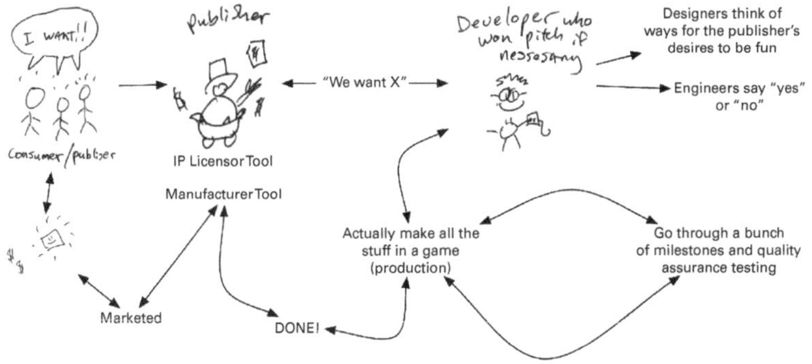

Figure 6.1
An artist's interpretation of the Actor-Inter/Intranetworks

AUTHOR_DEV_DILEMMA: The game industry is a favorite topic of discussion for game developers. More than any other subject, it is what my informants were the most interested in discussing. The day-to-day work of game development was never as interesting as thinking about the structure of the world that supports their labor.
Casey: Talk to me a bit more by what you mean when you say "the industry."
ART_Spidey_2: Ok, so there are a bunch of people. There are the consumers; they have a need, or a want. Now, whether they are out there saying, "We want a game," or if the publishers are saying, "We think they want this game," either way, the next person is the publisher. Basically, publishers are the people with the money.

(Continued)

Casey: Did that not use to be the case?
ART_Spidey_2: Well, it seems that they've always had money. But I guess even those places started as spin-offs from Atari and places like that. But they have it now, so they use it to make more money. So, pretty much, the publisher says, "We think making this game will make us money," and either they look for an internal or external developer. So, Activision has multiple internal developers, who sometimes even compete with one another. But, the publisher goes to the developer and says, "We want X," where X is a cool game they think will make money. So sometimes developers have to fight over that and pitch ideas to the publisher. Eventually they'll decide on one and make a deal with the developer. But no matter what, the developer has to figure out how to make what the publisher wants, but make it fun. Publishers could really care less if it's fun, as long as it makes money. Now, sometimes there is a licenser involved. . .
Casey: Like Marvel or DC?
ART_Spidey_2: Exactly. Now, sometimes the licenser is the publisher. Other times it isn't. But, the publisher has already negotiated that part of the deal. But, there is always this big kind of loop or tug of war, where you're coming up with ideas . . . Oh, and the engineers are in on this tug of war too, along with your designers . . . But the designers are saying what they want to do, and the publishers and the licensers are saying what you're allowed to do, and the engineers are saying what they think they can do. It's a giant tug of war that can fester for a long time sometimes. But you eventually wind up with an agreement and no one is really happy, and then you start making the game.
AUTHOR_DEV_DILEMMA: For many, the interest in the industry, or the ability to speak knowledgably about its structure and critique those elements that they find frustrating is akin to a rite of passage. A common sentiment among my informants was: if a developer isn't at least a bit cynical about the industry, then likely that developer hasn't been working long enough. The more entrenched a project is amongst these networks of licensing and publishing, the more likely a developer was to have thoughts about that structure. Those developers who by some stroke of luck began working on an original IP often take longer to develop their more jaded set of understandings of their position in the game industry.
#: SET DEMO_MODE 0

World 6-1: Why the Console Face?

"The industry" is an object of scrutiny for nearly all game developers. It both constrains and compels them; it is the broader system within which they play. The image that begins this world's DEMO_MODE nicely demonstrates how many developers feel they fit into the networks of videogame industry. Though developers are quite close to the creation of the product, they end up feeling quite distant from the consumer, the teams of marketing specialists who decide how to sell the game, and the money, which ultimately funds and fuels the videogame industry and their jobs. Many studio heads said that they consider themselves to be in a "service" industry rather than "product" industry because of their position in a network of companies. Relationships within the networks of game production end up mattering as much as (or more than) the individual projects that are developed. This in many respects counters the belief that the product focus of the game industry is one of the features that differentiate it from other new economy workplaces. The overarching argument for World 6 is that, as the industry has "matured," the networks have become less accessible and less interoperable. This trajectory consequentially limits developers more than they might like to believe (and in ways many are unwilling to criticize).

The structuring effect of the network is particularly interesting. Network approaches to understanding work, the economy, production, or society frequently fail to actively engage with the structuring effects, or more generally "power" in a very un-theorized sense. The "flows" of knowledge, which are then networked, are also structured in ways not addressed in current research (Castells 1998). "Access" is such a key aspect of the game industry and of game development work more generally, yet it is frequently glossed over in research that attempts to examine the networks of game production. These studies rarely look closely at what is necessary for a developer to gain access to these networks (Johns 2006). The concept of the inter/intranetwork is a useful tool for thinking about the structuring effects of networks. The structure that has emerged is "networked" but, more explicitly, networked in a fashion that I have termed "intranetworked," or closed off. Much like a corporation's private internal network, or intranet, the highly networked structure and knowledge flows are tightly controlled and connections to the broader world or "internetworks" are highly monitored. Many Internet users imagine that they are "on the Internet" while accessing it via Internet Service Providers (ISPs) (such as Comcast, Verizon, AT&T, or America Online). In fact, this is largely not the case. Most ISPs actually operate as intranetworks that provide access to the Internet through controlled mechanisms.

Box 6.2

#: SET DEMO_MODE 1
AUTHOR_DEV_DILEMMA: Game developers, while quick to reflect on their position within larger networks of production, very rarely attempt to critique publicly those structures. The Game Developers Conference is one space where those criticisms are leveled, yet once developers leave that space, rarely is that critique maintained and explored fully. It seems as if GDC operates as a release valve that allows excess steam and pressure to escape, yet sustained critical analysis of those structures are rarely carried on.
Casey: Do you think there is something about game developers that encourages them to be more self-reflective?
STUDIO_CREATIVE_HEAD: That's an interesting observation. I think it is true of the game development community more than the industry at large. I hate to say it this way, but, I think if it were up to most game publishers or console manufacturers, there really wouldn't be a Game Developers Conference. I'm not saying it's an "us versus them" sort of thing, just that game developers are often more forthcoming with information and knowledge, giving away secrets, and generally being self-deprecating, programmers especially, that doesn't sit well with the big companies. But, things are changing at such a rate that keeping that information secret is less helpful to you than sharing more. And, the industry is less open than it used to be. It used to be a lot of black magic, hacker in the bedroom sort of thing, so it made sense to not share your competitive advantage. But things are just so big and so dynamic, and the only way we're going to learn is if there is a lot of active openness and sharing.

So, that is there, and the culture has evolved, which is cool. You see a lot of public postmortems and people sharing with one another at GDC. And you really need it, because things are moving quickly. It used to be that cutting edge stuff took ten to twelve years to make it into a game, from the academic world to the industry. Now you'll see something at SIGGRAPH in the summer and then a hacked up version of it at GDC the following spring and our guys are figuring out how to incorporate it into our engine a week later. Then you'll be doing that on hardware that isn't finished and finding bugs in Microsoft's compiler as they're trying to release the new hardware. Developers are just trying to adapt themselves to this rate of change that is really fast as well.
AUTHOR_DEV_DILEMMA: I've most frequently characterized the lack of sustained analysis of the structure of the game industry to be a product of "Ooh, look, shiny!" Developers are so frequently enamored by new tools, tricks and technologies that they forget the critiques that were being made at any given moment. As the

```
game industry chugs forward, there is always something new to
explore or hacked that enables a new world of exploration. Devel-
opers are so interested in and passionate about their craft that
it becomes difficult to maintain a critical examination of its
context, because that isn't as interesting as making another game.
#: SET DEMO_MODE 0
```

The concept of intranetworks that structure themselves in ways that close off knowledge is useful in the anthropology and sociology of science. These disciplines use "actor-networks" as a means to analytically understand how science and scientific practice unfolds. While actor-networks provide some insight into the gaming industry, I have become critically interested in why particular nodes become obligatory passage points, or why entire networks become closed off from other nodes in the network: "If technoscience may be described as being so powerful and yet so small, so concentrated and so dilute, it means it has the characteristics of a network. The word network indicates that resources are concentrated in a few places—the knots and the nodes—which are connected with one another—the links and the messages—these connections transform the scattered resources into a net that may seem to extend everywhere" (Latour 1987, 180).

This game industry has become highly structured with very little intention or forethought toward structure. The networks of the videogame industry are inter/intranetworked and may seem to extend everywhere, but are accessible to only specific individuals and organizations. Publishers consolidate their interests by acquiring smaller (moderately) successful game development studios. Console manufacturers (who are also frequently publishers) do this as well. Frequently, new connections end up closing off networks. This often results in metaphorical islands of information disconnected from the mainland. Even independent studios tend to operate only in concert with a small number of other studios if any at all. This perpetuates the structures of inaccess (especially for young, small game companies) and results in islands of game production practice with many developers and studios remaining completely disconnected from broader structures that might enable the industry to mature.

The inter/intranetwork structure stands in stark contrast to how networks are frequently talked about, particularly in the context of the new economy. Sociological inquiries into new economy work has drawn heavily upon the network metaphor, emphasizing their limitlessness over their structuring effects: "Networks are open structures, able to expand without limits, integrating new nodes as long as they are able to communicate within the network, namely as long as they share the same communication codes" (Castells 1998, 501).

But as I've already established, the communication codes in the videogame industry are largely closed, and must be rediscovered by many aspiring developers looking to enter the videogame industry. This lack of openness and collaboration is fostered by the highly restrictive legal agreements and sense of secrecy that dominate the videogame industry. Thus the game industry (in)effectively attempts to maintain this kind of complex corporate, social, technical, and legal network.

Videogame development companies—studios—are where games are created. It is within these companies that code is written, art is generated, and designs are made. The communities of game developers are not that much removed from those of other artistic networks of production. "Information circulates through networks: networks between companies, networks within companies, personal networks, and computer networks" (Castells 1998, 177). Sociological analysis of artistic networks demonstrates compelling parallels to the networks of the videogame industry. What seems crucially important, however, is that artistic networks, despite their scale, rely on other networks of production.

> Some networks are large, complicated, and specifically devoted to the production of works of the kind we are investigating as their main activity. Smaller ones may have only a few of the specialized personnel characteristics of the larger, more elaborate ones. In the limiting case, the world consists only of the person making the work, who relies on materials and other resources provided by others who neither intend to cooperate in the production of that work nor know they are doing so. (Becker 1984, 37)

In much the same way, there are several different kinds of game development studios. The most basic and most nebulous is the independent, or self-funded studio. They range in size from several developers, artists, and designers working together to create a videogame to large companies without an exclusive relationship with a particular publishing company or console manufacturer. These companies or loose affiliations of individuals create games of their own design.

Most independent developers eventually enter into some kind of relationship with a publisher to release their first game on a PC. Then they must begin responding to the desires of the publishing company to ensure the release and distribution of their game. If the game is successful or garners critical acclaim, then the relationship with a publisher continues, and only then does a game development studio gain access to the resources necessary to develop console games. In some cases independent developers instead opt to distribute their games online as a downloadable game for the PC or via Adobe Inc.'s "Flash" Web-based technologies. Either way, independent developers take on the task of distributing and supporting their game, which is not the most lucrative or prestigious path in the videogame industry. Often game developers consider this path amateurish or as something that one does before you have actually "made it" in the industry.

Once independent developers have proven their ability to successfully create and release a game, they frequently become what is called a "third-party development company." In some cases companies will enter this category as a means to make money during the development of their first independent title. Other companies enter this phase immediately, if they are created by developers with connections to publishing companies or console makers who have diverged from other game companies. A third-party developer is similar to an outsourcer for a publisher; they are instructed to make a particular game. In some cases a previous game title can be used as a reference. This is most common in the case of established franchises. For example, a company may be contracted to develop "Shrek 3" or "Batman Forever." Developers will frequently play or examine the titles that came before it. In most cases, the publisher already owns, or has acquired the rights to these "IPs." The developer is then authorized to make a game to the publishing company's specifications. There is often not a clear delineation between a third-party developer and an independent developer. Most third-party developers have internal, independent projects, which are funded by the revenue based on "non-independent" work for a publisher.

Box 6.3

#: SET DEMO_MODE 1
AUTHOR_DEV_DILEMMA: One thing that all developers I spoke with seemed to believe is that the game industry moves in very cyclical ways. That as conservatism spreads through the industry, new risk is taken on at the margins and new innovations then break free, allowing for new forms of creative exploration. In the meantime, however, there are many cases where developers feel as if they are put into situations where the limitations placed on creativity will ultimately result in failures that could have been avoided if those limits had not been put in place.
Casey: What is the worst part about making games, then?
ART_Spidey_2: It has to be that it is difficult to make imagination fit into a process. But it also has to be a viable business. When we come up with a really cool idea that we think will be

(Continued)

fun, that isn't the only thing that determines if it goes into the game. It has to be approved by whoever is licensing the character. It has to be compared with what marketing thinks is popular that day. They have to promote things in games that may not be fun, but they think increases sales. It really is the relationship between the suits and the beatniks. Sometimes dealing creatively with restrictions is cool, but when it's just seen as an outlet for sales rather than a creative thing, that isn't fun.
Casey: Do you think as the industry pushes forward, pushing that imagination into a process, that things will be more structured?
ART_Spidey_2: I think it will go back and forth. I like to compare it to movies. Movies probably started the same way. There are people with crazy ideas, who know people with money, but those producers say, "Hey, you know that idea is pretty cool but I don't see it making money. I can give you $300 million to make this other thing though." So you reach a point where you're making things, like "White Chicks," and then you have a resurgence of people with crazy ideas. Videogames do the same things. Videogames have become so expensive to make, which is why Nintendo has really tried to change the direction of things. Nintendo knows that small developers don't have $40 million to make a game, but would rather them make something weird like Nintendogs or Katamari Damacy. I always think it will be a tug of war.
AUTHOR_DEV_DILEMMA: Restrictions were always complex for my informants. In some cases limitations were seen as creative opportunities, while in other cases they were viewed as distractions. The distinguishing factor seemed to be linked more to the origin of the limitation. Hardware and software limitations are fun and interesting, while restrictions made by those seemingly disconnected from the creative process are viewed less positively.
#: SET DEMO_MODE 0

While it may sound as though third-party developers are simply "outsourcing" houses for publishing companies, this would misrepresent the dynamic and complex relationship between developers and their publishers. Developers are not given a precise description of the game, though they often want one. Instead developers must frequently base new designs on older ones, which are then vetted by the publisher. The oversight of publishing companies over third parties is varied and complex. In some cases very little direction will be given except at milestones or intermediate steps along the development process. In other situations, publishers may place a producer in the offices of the third-party developer to provide constant feedback. Frequently publishers have a conceptual foundation that third parties can begin working from, but the actual game that winds up on the shelf is a product of both companies working in concert. This means that publishing companies in many cases do little to zero actual game development. The design "document" that becomes a part of the "contract" is actually developed by the game studio. The effort of creating games is reserved for those working for game studios.

The distinction between outsourcing and third-party development is important to make, considering that there are true outsourcing companies in the videogame industry. These may also be individual freelancers who create art assets, localize text, port code to new platforms, or test games for defects. These relationships are governed with relatively precise specifications and contractual obligations on both ends. Many game companies in India have chosen to use outsourcing as a means to fund internal development projects. While most game companies in the United States begin as a mixture of independent and third-party studios, most Indian companies begin as a mixture of independent and outsourcing studios. In part this has been due to the readily available manpower with experience using and creating media arts. Since code is frequently highly protected by game studios, they contract for very little code outsourcing. Contractual obligations in many cases govern whether or not an outsourced studio is even listed in a game's credits, which are critical resources for aspiring game companies, as bylines begin to establish credibility for a game development studio. However, given the climate surrounding outsourcing in the United States, companies will often pay more to prevent outsourcing studios from speaking about or placing their logo in games bound for the United States, for fear of consumer retribution or bad press. As previously discussed, this restricts employees at these studios from being able to claim having worked on a title, which limits developers from gaining access to social networks within the game industry.

There is a typology among those companies doing offshore outsourcing work within the videogame industry. This differentiation is primarily based on the markets that a company is most interested in venturing into. For some, the drive is developing games for an internal market, closely targeted to users in their home countries. For others, it is the eventual creation of games targeted at a global market, much like their counterparts in already established markets. Other companies seem to be interested primarily in acquisition by large multinational publishing companies (which is not to say that they do not have aspirations of those other companies, simply that they have charted a different course). Such goals are realistic, since recently, publishing companies have begun acquiring outsourcing studios in India, China, Vietnam, and other countries, hoping to use these acquired studios as sites for intracorporate off-shoring. In these cases, publishing companies have established production pipelines for particular games, which only need content created for them. These acquired outsourcing studios are then used to create content in a highly controlled but cheaper environment.

Closely related to the outsourcing companies are "middle-ware" companies, which provide software that enables more rapid development of game systems. As the complexity of games has risen, this new class of company in the videogame industry has exploded, and in many cases these companies have sprung from countries that otherwise do not have a large established videogame industry. Rather than outsourcing, these companies have developed extensive libraries of source code, software tools, and process management systems, which they can sell to developers in Japan, the United States, and Western Europe.

The final kind of company in this typology is the in-house development studio, which is wholly owned by either a publishing company or console manufacturer (who are often publishers as well). These studios frequently act like independent developers, third-party developers, or a mixture of the two, not unlike studios not owned by publishing companies. There is often minimal collaboration between studios under the same publisher. Different studios have different practices, systems, technologies, processes, and internal cultures. In rare cases collaboration does occur, though in most cases the extent of in-house-development interaction with publisher or console manufacturer is through the interface of studio heads and employees of the parent publishing company.

Box 6.4

#: SET DEMO_MODE 1
AUTHOR_DEV_DILEMMA: Consolidation within the game industry was an oft-cited concern that centered frequently on creativity and the kinds of games being produced. More and more studios were being acquired in order to exploit the creative work those studios had already produced.
Casey: Are there aspects of the industry that make you nervous?
STUDIO_CREATIVE_HEAD: Yeah. There is a lot of consolidation in the industry, which occurs for different reasons. When that happens, people fundamentally stop prizing the best, because you want to see growth. But, you can do it organically too. Mostly it seems to be about economies of scale and being able to take on

```
risk and sustain taking a hit, if a product is a giant bomb or
something bad happens. There are all those reasons for consolida-
tion. Game development and the game industry at large is inher-
ently a very risky business, and things are getting more risky,
not less. There are bigger costs than ever before, so mistakes
are very costly. You are always making a lot of bets and if we're
making a big bet, we're going to play it safe. Sequels and other
things are safe bets, you know they'll at least be commercial
successes. Unfortunately, that is counter to innovation and you
have stagnation and that stagnation is what kills the industry.
So it's kind of a self-fulfilling prophecy. What happens when the
safe bet is actually a disaster?
AUTHOR_DEV_DILEMMA: Many developers point to stagnation as being
one of the defining characteristics of tectonic shifts in the
game industry. Major shifts came from companies taking bigger or
more frequent risks, not from fewer.
#: SET DEMO_MODE 0
```

For many developers in India, the game industry is very distant. In some locations developers have mobilized through informal meetings, or through local International Game Developers Association (IGDA) chapters. In other areas companies have banded together with larger organizations, like India's National Association of Software and Service Companies (NASSCOM) to encourage new growth in game development. Of course, for some, like those working for Microsoft's Casual Game Group in Hyderabad, India, whose games are being placed onto the online distribution network of Microsoft's Xbox 360 Xbox Live Arcade, the feeling of distance is minimal. The networks have already been established. For companies that have yet to gain access, the distance is palpable.

More than any other question I received while in India, I heard, "How does game development practice in India differ from what developers in the United States do?" This relatively simple question frequently led to conversations about a disconnect between what Indian developers are allowed to contribute to game development projects and those tasks that are necessary to produce a videogame from start to finish. Some companies do create games from start to finish, though at a different scale. They create games for mobile (cell-phone-based) game platforms. Although at a different scale, these networks are as difficult to access as US development networks are. Mobile game studios tend to fund their development efforts by also offering art asset production outsourcing services. Because of this, the companies become more specialized in one aspect of game development and don't develop their capabilities in others. This disconnect cuts deepest at the companies who become solely identified as locations for outsourcing by US companies. As the majority of these studios' resources become focused on that singular aspect of the development process, the studios become disconnected from game development more broadly.

By contrasting game developers with the workers in other technologies we can see how isolating outsourcing one facet of games can be. Science and Technology Studies scholars have demonstrated how Indian scientists and engineers have remained in conversation with American institutions through electronic means.

> [India's] scientists and engineers are highly connected with their peers in American institutions. This is partly because scientists and engineers in India overwhelmingly enjoy access to the World Wide Web, but institutional linkages are even more important. With the IT revolution, Indian S&E educational institutions have been increasingly connected with the United States, as well as with the rest of the world, essentially comprising one global system. (Varma 2006, 40–41)

However, this interconnectedness has largely not been the case for game developers. Scientists and engineers, unlike game developers, have avenues or venues in which collective knowledge is shared more broadly. Here, again, game-industry secrecy prevents the formation of a community of practice more broadly at the cost of limiting the growth of developers and the industry.

For those developing-market developers interested in projects for any platform outside of the Web, personal computer, or mobile (and even this platform is notoriously difficult to work with due to the domineering attitude of carrier companies), the opportunities are extremely limited. It is frequently only when the concerns of a US company are involved that some sort of agreement can be reached and the requisite hardware, software, and documentation are finally given to Indian developers. However, this information is often provided without connection to the tacit knowledge of what it takes to create a game for these platforms. The immense body of knowledge that has become codified only in the practices and conversations of developers is not transferred along with the capabilities to produce games for these systems. When Indian developers go through the same learning process that other established developers went through only several years earlier, and which could be ameliorated with access to existing systems and knowledge, Indian developers are confronted with questions from experienced game developers that amount to "Don't you know anything?"

This network disconnect is not simply limited to those in distant countries. Numerous independent developers, and even those simply struggling to bring developers together away from the United States' East and West Coasts face similar barriers. As mentioned before, if you are not part of the game development community, the only way you can get in is to create a game, but it can be difficult to develop a game without connection to those existing networks. Instead, you must fumble your way until you have learned enough on your own to prove your worth, at which time you are best served by moving to where the networks have already been established.

As World 2 discusses, there are secret social networks as well. "An industry's cocktail parties, seminars, and informal gatherings form its social backbone and are especially important to innovative industries that rely on the rapid dissemination of information" (Neff 2005, 135). In the case of the game industry the glaring gaps are less about the dissemination of rapidly changing information, and are more about social networks. These closed intranetworked social structures "increase the experience of labor market inequality" and "workers unable to access or maintain these networks may be at a disadvantage" (Neff 2005, 138). These closed networks are not just pervasive, but assumed normal and necessary. Most contradictory is that workers must negotiate access to these networks and simultaneously maintain the barriers to entry.

The social inter/intranetworks of game developers are not without differential power relations. In the move from despotic power relationships, like those of manufacturer/publisher/developer, to hegemonic relations, more akin to those of core/periphery, a new kind of measure, one of connectivity and reputation, seems to be ever more predominant. This new power relationship concurrently attempts to obscure the labor of rank and file developers. This is not an isolated activity, however, since anthropological analysis of social networks among high-energy physicists demonstrate similar boundary marking and maintenance by practitioners of scientific work. Knowledge work and creativity combine in a highly competitive space for funding that results in very similar kinds of border preservation.

> Networks of exchange link otherwise autonomous units at every level of social organization. The primary commodities exchanged are students, postdoctoral research associates, and 'gossip' (oral information about detectors, proposals, data, organization of groups and labs, and the location and professional genealogies of individuals). The boundaries of the networks as a whole are closed, marking off the outsiders. . . . The boundaries of the community as a whole are negotiated with great circumspection. (Traweek 1988, 123)

The use of reputation networks as a mechanism for structuring numerous resources within a section of networks is consequential for any social network. The same is true for labor and knowledge production networks. If the practices of a given subsection of the inter/intranetwork are not meeting the expectations of other components of the network, their reputation, and subsequently their income will begin to fall. While these networks are social, they are also technological, corporate, and intricately connected to complex legal and legislative systems. In effect, the network structure has systematically blocked out those mechanisms by which access for developers both foreign and domestic can be granted. More and more work is being done only from within the networks. Those hoping to break into the network must battle numerous difficulties in what is largely (and falsely) being touted as a "flat" economic system.

World 6-2: The License

Box 6.5

#: SET DEMO_MODE 1
AUTHOR_DEV_DILEMMA: The current model of licensing that dominates the game industry is not an accident, nor was it inevitable. It was designed with specific goals and interests in mind with little consideration for what those might mean for game developers and the craft of game development.
Casey: What about the relationship between the publishers and the console manufacturers?
ART_Spidey_2: There has to be money under the table, just to get games out, but I don't really know that. I mean, the manufacturer doesn't really care if a game does well or not. It makes money if a game does or doesn't sell. They make more if it sells well though. But the publisher is paying for the cost of production. I mean, if you wanted to sell a game that just crashed, and were willing to pay for its production, I don't think the manufacturers would really care.
Casey: But licensing started with the Seal of Quality, right?
ART_Spidey_2: Yeah, that was the whole Seal of Quality thing. You do need the manufacturer to provide you the hardware for you to make the game too. So they do get final say. Maybe that made sense when there were fewer games for each system, and if a bad game came out for a system people were like, "Nintendo is an inferior product." But that really isn't the case anymore. Nintendo used to run games through "Mario Club" after lock-check. It was sort of like quality assurance. Now they just run it through a little profiler to see if it runs all right and doesn't crash. There are some legal requirements and things about making sure when a user does something like unplug a controller it says a

```
very specific thing. But other than that, they don't care much,
at least it seems to me.
AUTHOR_DEV_DILEMMA: While quality may have been the primary con-
cern of licensing in the early days of the game industry, things
have changed in the subsequent 30 years. The logics of licensing
have changed and the necessity of it has also shifted.
#: SET DEMO_MODE 0
```

You can always tell when an executive producer from one of the publishers or manufacturers is "in the house" at a game studio. Parts of the company seem to exude a new level of stress and excitement. Secret meetings are organized among select individuals (often in buildings made with more glass than walls, which results in mobs outside the glass observing a small team of workers within). The visit often denotes a demo of new hardware for a development team expected to come to grips with it over the coming months to add it to their supported platforms. Or perhaps a new build of a game is being shown to receive feedback from the producer. Regardless, it is almost as if the emperor or one of his inner circle has traveled from afar to see what the commoners have been preparing in tribute.

As far as developers are concerned, publishers and manufacturers control the videogame industry. Development studios structure themselves around their relationships with these companies. As already noted, developers can create games for the PC and the Web without publishers and console manufacturers. Why is it then that so much emphasis is placed on the ability to work with publishing companies on console videogame systems, aside from the prestige of being allowed to do so? What is it about consoles that separate them from other gaming systems like the common PC? Console gaming systems did not mark the beginning of computer-based gaming, a point made by numerous studies of the "birth" of the videogame (Kent 2001; Kline, Dyer-Witherford, and de Peuter 2005; Malliet and Zimmerman 2005). But consoles did several things differently from the (personal) computer, the first being the way that people interacted with the system. Typically this was through a simplified manner of data entry, perhaps by a rotating knob and button, a joystick with one or more buttons, or with something like the controller of the Nintendo Entertainment System with its directional pad (visually similar to an equilateral cross) and one or more buttons. Consoles were connected to televisions rather than to separate and costly monitors or "dumb" terminals connected to mainframe computers. The console was simply another component of the growing "entertainment center" in the home.

Table 6.1
VGChartz sales data 2006–2010 (Staff 2011)

Year	Console game sales	PC game sales
2006:	231.8 million units	1.2 million units
2007:	388.4 million units	5.2 million units
2008:	628.5 million units	9.3 million units
2009:	602.3 million units	8.8 million units
2010:	636 million units	17.6 million units

Price is a major issue in where the industry focuses its development power because consoles are significantly less expensive than the computer configurations necessary to play games, which means more customers have consoles than computers. The average purchaser can more easily justify several hundred dollars for a console rather than several thousand for a computer. As time went on and the price of personal computers began to fall, more people acquired them. Many believed that console gaming was something that would vanish, a remnant of history (Carless 2007; Edery 2007; Snow 2007). On the contrary, console gaming continues to be most lucrative sector of the videogame industry. Simply by looking at the differential scores between PC and console game sales (see table 6.1), you get an idea of the magnitude of difference between the sectors. This would explain publishers' interest in the console systems.

This differential is only partially explained by the simplicity and lower cost of gaming consoles. The more significant difference lies in both the capabilities and limitations of a gaming system. Consoles frequently have certain technological capabilities beyond those found in affordable PCs. In the beginning, graphics processing power marked the difference. In many cases the graphics processing power of console gaming systems when initially released have been beyond those of the average computer. Consoles begin their lives more capable than average computer systems even though throughout the lifetime of a console similar technologies frequently become incorporated into general computer systems. Further, until recently, standard operation for console makers was to price consoles at a loss. The basic logic of this practice is that console manufacturers make more money from games than from hardware. The console system itself was historically a loss leader because technologies were so much at the cutting edge that they were offered to buyers at prices below cost. Not surprisingly, this brought about the downfall of more than one console manufacturer.

Another reason consoles tend to outsell PCs centers on game production limitations. Because they are not required to do "typical" personal computing tasks, consoles can be built with very precise specifications and limitations to their possible uses. From a game development standpoint, there are no differences to be taken into account with a console—no differences in amount of memory, graphics systems, operating systems, processing power. There are no warnings from a console that you do not have the proper driver installed. Each console is created with specific specifications, and when creating games for it, game developers know these capabilities. The case is very different on the PC. Differences in software/hardware configuration, the possibility of other processes running and interfering with the functionality of a running game, and simply the unknown availability of options frequently provides great difficulty for companies making games for the PC. Some argue, and I tend to agree, that these "non-standardized advances in home computers," have encouraged the growth of the console videogame industry and massive improvements in the graphics processing capacity of videogames. These developments also create one of the biggest worries and limitations for developers hoping to create games for personal computers (Williams 2002).

The focus on console games by developers and the perception that it is the only market worth developing for (or is the most prestigious market to develop for) has meant that consoles remain a focus for most developers. However, unlike personal computers, you cannot simply begin developing games for a console without first establishing a relationship with a console manufacturer. Because of this, these companies continue to have the single largest impact on the videogame industry.

The rules of game development for consoles are dominated by legal, technological, and monetary rules dictated by console manufacturers. While most developers avoid the pitfalls of legal prosecution, some do come under legal scrutiny if they attempt to circumvent the technological limitations of their position. The technological collapses into the legal in this space, as technology is often used to enforce already agreed upon legally binding rules. NDAs and contracts are secondarily enforced by technologies (DevKits and SDKs) leased to developers that enable their work. Yet, developers do have the ability to push back against console manufacturers in these agreements but do not for fear of losing access to the closed networks of production. However, it is precisely this commitment to broader sharing and collaboration that are crucial to the long-term health of game development practice. While technological limitations would still remain, it would be a step in the right direction. Developers could have the most impact in their desire to both shape and care for the industry if they used their ability to push back against the restrictions of manufacturers. At the same time, this restrictive part of the game industry disappears from their perception. Once a company has gained access to DevKits and SDKs, these tools recede into the background of a company, despite the fact that they were once one of the major gatekeepers of access to industry networks. Far too quickly developers allow themselves to forget just how difficult it can be to work into and among the structures of the industry.

The situation is even more complex though, because frequently before a request for a DevKit is granted, developers must demonstrate a proof of concept or playable demo of a game. This is frequently done using freely available resources and tools, targeting the personal computer or the Web. All of this work must typically be redone once a DevKit is acquired, as those systems are completely different from those freely available. This is a particularly sticky aspect when talking about a business rather than a hobby or something set up in your garage in the hopes of making it big. Many informants noted a distinct difference between game development as work and game development as play, especially when livelihoods are on the line. The shift is part cultural, with the activity of making games for oneself rather than making games for a pay-check being very different tasks. But there is also difference in the material practice. Tools and professionalization is linked more distinctly to sustainable game development work. Game development outside of those lines need not be sustainable, for it isn't a hobbyist's goal or purpose.

Even if a game development team is successful creating a game for console systems (enabled by leased DevKits) for distribution on the proprietary Nintendo, Microsoft, or Sony networks, the resulting game will require approval of the self-governing ratings system of the game industry, which will cost another $2,000 to $3,000. There is a chance that a game might receive a rating from the Entertainment Software Rating Board (ESRB) that limits the audience too significantly and requires massive rework. This creates a logic that leads many developers to ask publishers to fund the massive effort required to make a game, though that investment comes at the cost of relinquishing some control of the project, and frequently the rights to the IP of a project to the publisher. This way if the risk does pay off for the publisher, they can continue to reap the benefits of that adventure capital for years to come.

Despite all of this, risk and the rules by which those risks are taken remain a largely invisible aspect of the game industry. While developers will frequently lament the way the industry game is played, they seem blind to the rules by which it is played, something uncharacteristic from people (self) trained to pay attention to underlying structures.

Let's take a major industry shift as an example. Nintendo first emerged as a force to be reckoned with during the time of Atari, and the NES release was particularly important because it brought about massive change to the way videogames were developed, sold, and marketed—in fact, NES brought changes to the very technological core of game consoles. For the most part, these changes were not obvious to the user. The only visible difference was the emergence of the Nintendo Seal of Quality on the boxes of games. It was placed on "official" game titles released for the NES. These were the games licensed by Nintendo.

Nintendo's logic was clear. They believed that the low quality of games released for the Atari were partially to blame for the "crash" that came shortly after the release of several games that were massive economic failures. By offering their "guarantee" of quality, Nintendo offered gamers a way to feel safe investing their money in the games.

But this was not simply a quality control issue (if it was you would have likely had two groups of games for the NES, those "guaranteed" and those that were not). There were very few "unlicensed" games for the NES. The only games available (for the most part) bore the Nintendo Seal of Quality because what lies behind the Seal of Quality endures today on all games released for Nintendo's current generation of consoles. The very term "licensing" entered the minds and vocabulary of game developers at this significant moment in gaming. The real power of the seal is in the resulting web of connections and network. But why focus on manufacturers like Nintendo? It is because they structure the network more so than other actors. This is where concepts like those proposed by sociologists examining the "network society" or "networked/new economy," rightly index the importance of switches or points of control within the network. Unfortunately these important aspects are frequently not examined closely enough, despite their ability to dramatically shape resulting networks: "Switches connecting the networks are the privileged instruments of power. Thus, the switchers are the power-holders. Since networks are multiple, the interoperating codes and switches between networks become the fundamental sources in shaping, guiding, and misguiding societies" (Castells 1998, 502).

The Seal of Quality offered such a network switch, because of the ways that these networks structure the very work of producing games. Reading into several court cases, we can learn more about Nintendo's licensing practices, which are otherwise invisible. Up until the introduction of the

NES, companies were created games without licensing. The Seal of Quality changed all that. But what kind of deal was necessary to change the rules of the industry game? Atari's frustration at having the rules switched mid-game is evidenced in their antitrust suit against Nintendo (Atari, Tengen, and Nintendo 1992).

In December 1987, Atari became a Nintendo licensee. Atari paid Nintendo to gain access to the NES for its videogames. The license terms, however, strictly controlled Atari's access to Nintendo's technology, including the 10NES program. Under the license, Nintendo would take Atari's games, place them in cartridges containing the 10NES program, and resell them to Atari. Atari could then market the games to NES owners. Nintendo limited all licensees, including Atari, to five new NES games per year. The Nintendo license also prohibited Atari from licensing NES games to other home videogame systems for two years from Atari's first sale of the game. (Atari, Tengen, and Nintendo 1992)

Five games per year, and all costs must be paid for at manufacture. A company's entire earnings were limited to five games per year, and those companies bore all of the risk associated with the costs of production. Heaping limitation on top of limitation, those games had to be maintained as NES exclusives for two years before they could be ported to other console systems. Nintendo was the only company unhindered by these limitations on production. If other developers or console manufacturers attempted to change the rules, they met with not only the ire of Nintendo, but also the force of the state apparatus. The legal ramifications of copyright and patent systems were leveraged by Nintendo to alter the entire playing field of the videogame industry. But what did it cost other companies to play on this field? Was it fair to have different rules for Nintendo, since it needed to cover its expenses in the manufacturing of games for their console? Atari lost this particular case and was forced to play by the rules dictated by Nintendo.

Things have changed, and Nintendo no longer places such severe restrictions on the number of games a publisher can create in a year. This is as evidenced by looking at the number of games released for consoles each year by different publishers. But, if we examine the top publishers of videogames (see table 6.2), an interesting trend emerges.

No matter the year, every single console manufacturer is in the top ten. The ability to control what makes it into the content stream obviously has effects on who is making money. A particularly interesting outlier is Electronic Arts, who year after year manages to displace even console manufacturers. The small amount of motion you see in charts like this shows that only very large and very established publishing companies are managing

Table 6.2
Top ten videogame publishers in 2004–2009 (Wilson 2008; Staff 2009)

Top videogame publishers					
2004	2005	2006	2007	2008	2009
Electronic Arts	Electronic Arts	Electronic Arts	Nintendo	Nintendo	Nintendo
Microsoft	Activision	Nintendo	Electronic Arts	Electronic Arts	Electronic Arts
Sony Computer Entertainment	Microsoft	Activision	Activision	Activision	Activision Blizzard
THQ	Nintendo	Sony Computer Entertainment	Ubisoft	Ubisoft	Ubisoft
Ubisoft	Sony Computer Entertainment	Take Two	THQ	Sony Computer Entertainment	Take Two
SCi/Eidos	Ubisoft	Microsoft	Take Two	Take Two	Sony Computer Entertainment
Activision	Konami	THQ	Sega of America	Sega of America	Bethesda Softworks
Take Two	THQ	Ubisoft	Sony Computer Entertainment	THQ	THQ
Atari	Sega Sammy Holdings	Konami	Microsoft	Microsoft	Square Enix
Nintendo	Take Two	Sega Sammy Holdings	Eidos Interactive	Square Enix	Microsoft

to get their games into the console stream. The simplest answer is that this is an expensive game to play. Because the entire manufacturing run must be paid for in advance, and all marketing for a game must be covered by the publisher, creating games for consoles, while lucrative, is also extremely risky and requires a high initial investment. The game industry is built around high risk, high reward, high initial cost, and high barrier to entry rather than out of some innate necessity. This is precisely how the industry was engineered to function with the model implemented by Nintendo beginning in 1985. Yet the game industry need not function on such a model and developers need not labor under such a structure.

The industry's aversion to risk and high-cost power structure becomes crucially important when attempting to understand the relatively conservative behavior of publishers and the kinds of games that they are willing to fund. Unless they acquire a studio near the end of the game production cycle, publishers bear the majority of the risk associated with physically creating a game. This does not mean that publishers bear all of the monetary risk associated with game development, though many developers I spoke with believe this to be true. Frequently publishers only bear the full risk of a game development project if it is one that they have entirely sponsored the development of. In many cases these games already have proven franchises, brands, or more generically, IP (intellectual property, though in a fairly restricted sense of that term). Some have called this a move toward a "hit driven" industry, much like the movie industry, where for the major publishers to be successful every title must not simply be profitable, but massively so to recuperate the development costs. As one of my informants noted, this makes them incredibly risk averse, which is actually the opposite of what they need to be to truly be successful in a hit-driven industry. Publishing companies desire to play it safe means that they leech the profits of particular game franchises to death, rather than nurturing the kinds of environments where runaway hits can be fostered and grown.

This is compounded in the current industry situation, where the massive growth of available storage space on game media has caused many companies to place an emphasis on rapid expansion of game art assets. Many of my informants have pointed to modern games being "asset limited" rather than "engineering limited." Simply stated, the greatest percentage of a game's cost stems from creating content for the game, not the underlying code that puts a game into action. The release of "next generation" or "next gen" console systems has led many developers to focus on the creation of highly detailed art assets for games that frequently require more production time to create content.

This entire publishing and production picture is further complicated by the "unpredictability" of videogame development, which is frequently touted as being more difficult to manage than traditional software development, though some argue otherwise. Simply looking at the archives of *Game Developer* magazine's "Postmortems," you begin to see a pattern of unknowns coming back to bite game developers late in the production of their games. Frequently this results in rework for every aspect of the development team, where changes ripple across engineering, art, and design. This of course has serious repercussions when most publishing companies want their premier titles on retailers' shelves during rush buying seasons; in the United States this is the Christmas sales season. In other cases publishing companies have partnered with the movie industry using their established franchises to encourage sales of videogame titles. In each case this leads to a rigidity of release dates, and missing these dates can be disastrous for both publisher and developer.

To reduce these risks, as touched upon in World 5, publishing companies have begun offloading the risk of developing new IP or franchises to independent developers. While this is not possible for a publishing company hoping to make a game in concert with a movie studio, it is frequently the case for new games to come from otherwise unknown studios. Often the publisher becomes involved in the development of these games only after an independent developer has already developed large portions of a game concept, thereby capturing the adventure capital without the substantial risk. Once this is complete, publishers will frequently milk these new franchises with or without the original development team, depending upon the contract agreements between publisher and developer. In many cases if a developer retained the rights to their new IP the publisher instead purchases them so they can milk the new franchise regardless.

This mentality of capturing game equity regardless of the method leads publishers to have a certain conception of the market and consumers. In particular, because of the emphasis on "hits," most games attempt to capture the core or "hardcore" gamer market or existing, lucrative IPs. Franchises like "Barbie," "Bratz," or "Batman" are viewed as less risky than titles without an established market. "Madden NFL" or other sports franchises oscillate between being seen as a kind of bread and butter and being derided by game developers. While it is the job of publishers to identify and market games to consumers, many publishers have become even more conservative in their approaches, stonewalling development efforts that do not have an obvious brand or market associated with them. Despite this, once a game has proven its appeal as an online "Flash" game or downloadable game, many publishers will then take the substantially lower risk to further development and distribute the game via more traditional channels on more restricted hardware like console systems. In a sense, then, they use independent developers to test the market for them and then take advantage of the lower risk, proven games to reap adventure capital without much downside.

World 6-3: The Development Kit ("DevKit")

More than any other relationship, developers perceive their interactions with publishers as the most problematic. Even developers working for studios wholly owned by publishing companies view these as troubled marriages, in part because, the publisher is the gatekeeper to other networks in the game industry. But the relationship is further strained because the publishing companies exercise their position of power within industry networks frequently at the expense of developers' time and effort. The relationship between publishing companies and developers frequently take the foreground in informants' conversations about developer frustrations.

Obviously, as figure 6.2 indicates, developers seem to not be having the best time, perceiving themselves as the hard working talent lining the pockets of the slovenly game publishing pimps. Unfortunately for our

Figure 6.2
An artist's interpretation of the publisher/developer relationship

understanding and discussion of the industry, this relationship is often stereotyped and misunderstood because cultural analysts typically have only the insight of ea_spouse as their guide for what work is really like in the videogame industry. At the time the blog post and attenuated ea_spouse comment were published I was sitting with a group of developers working on a videogame based on an upcoming movie title for an unreleased handheld console. They, too, were in crunch mode, working to beat timelines that had been set to meet the demands of movie executives, game publishers, and console manufacturers, forced into a release calendar without realistic expectations about how long it takes to make a game. As we saw in World 1, the game was later canceled, but those hours late at work fighting against prerelease hardware with prerelease software development kits (SDKs), a new engine, an in-development build system, and no proven pipeline for art assets or design data were still fresh and raw for my informants. That is not to say that ea_spouse was wrong about the industry secret of pervasive crunch, but rather that the situation is even more difficult and complex than analysts had previously envisioned. Developers have not just traded their allegiance for televisions, but for a slew of new and interesting technologies and access to private networks.

As previously noted, one of these technologies, the NES, marks a pivotal moment in the history of videogame development, and in the realm of rule making and rule enforcement. In December of 1985, shortly after the release of the NES in the United States, Nintendo filed for a patent, the only public record of a significant technological device that had been developed.[1] This microprocessor, the piece of code carved in silicon, was about to change the course of videogame development forever; and in this case software/code presaged a code—a legislation—of another kind.[2]

The phrase Nintendo used in its patent application, "System for Determining Authenticity of an External Memory used in an Information Processing Apparatus," belies a much more complicated device, for within the console resided a semiconductor lock and in the cartridge a key. This silicon lock and key ensure the "authenticity" of the "external memory device" added technological force to the Seal of Quality placed on a game's box. The key and seal worked together to restructure the way games were made. To get a game made with the key you needed a seal. To get the seal you needed Nintendo. The following patent document excerpt (figure 6.3) demonstrates the "invention" of this silicon/digital lock and key.

Nintendo engineer Katsuya Nakagawa, in the device's patent, clearly lays out Nintendo's plan to have a locking mechanism in the console and an unlocking mechanism in the external memory device (game).

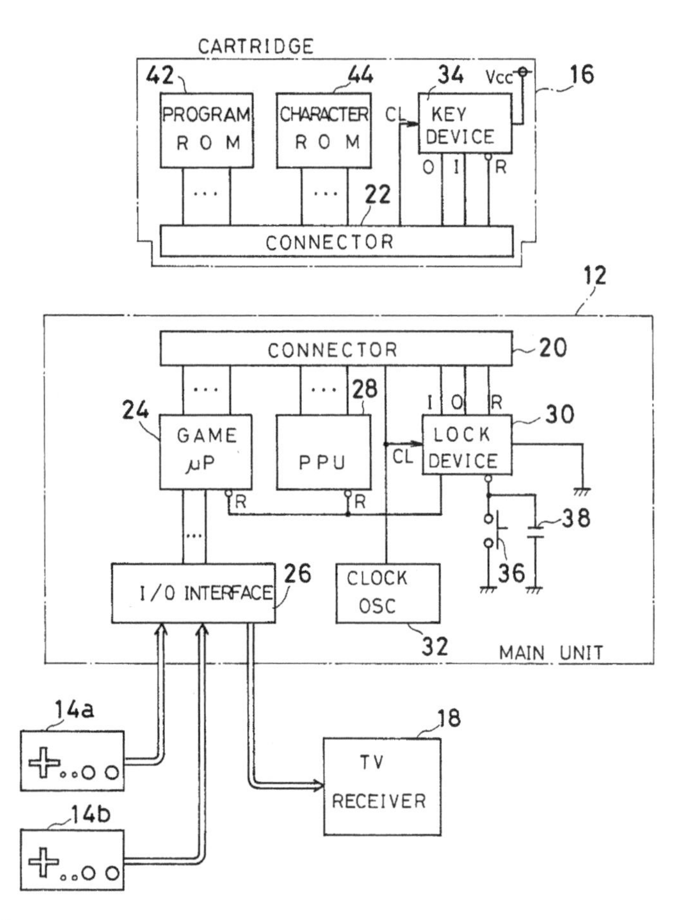

Figure 6.3
The lock and key of the 10NES patent

United States Patent Number 4,799,635 - System for Determining Authenticity of an External Memory used in an Information Processing Apparatus.
To verify that the external memory is authentic, duplicate semiconductor devices, for example microprocessors, are separately mounted with the external memory and in the main unit, respectively. The semiconductor associated with the external memory device acts as a key device and the duplicate device mounted in the main unit acts as a lock device. (Nakagawa 1985)

Nintendo's intentionality with regard to this method is broadcast in the court cases that quickly followed from those companies who did not care to work through the locked system wherein Nintendo locked out quality to lock in power. Nintendo's own testimony indicated that the 10NES chip, as it came to be known, was designed specifically as a means to enforce licensing agreements. Note that at this moment the interest in the way the lock and key were structured was in protecting Nintendo's ability to say who could make and release games for the NES, rather than on copy protection; and while I do not address those concerns in depth in this world, I will return to them in World 7-2 and 7-3. The intentionality of what the device was designed to do is directly indexed in a court case from 1992.

Nintendo designed a program—the 10NES—to prevent the NES from accepting unauthorized game cartridges. Both the NES console and authorized game cartridges contain microprocessors or chips programmed with the 10NES. The console contains a "master chip" or "lock." Authorized game cartridges contain a "slave chip" or "key." When a user inserts an authorized cartridge into a console, the slave chip in effect unlocks the console; the console detects a coded message and accepts the game cartridge. When a user inserts an unauthorized cartridge, the console detects no unlocking message and refuses to operate the cartridge. Nintendo's 10NES program thus controls access to the NES." (Atari, Tengen, and Nintendo 1992)

While Nintendo's technological legislation encouraging developers to work directly with Nintendo was dramatically different from how things had been managed for previous consoles, it was in many ways only the beginning. This leads to two questions: "Why not pick the lock?" which can be answered in part by the second, "Why the patent?" Nintendo had learned from Atari's experiences and knew that the kinds of people interested in making games were very resourceful. A simple technological device, while capable of influencing the way in which games were developed for the new NES, was not enough to ensure control over the rights of production. This power is simply out of the hands of most organizations. To exert that kind of control requires the mobilization of government intervention, which is precisely why they applied for a patent.

Interestingly, NES's Japanese counterpart, the Nintendo Famicom, did not contain the 10NES lockout chip, and while this did result in some levels of piracy that Nintendo combated with just the Seal of Quality (without technological or legal networks), the lack of 10NES also resulted in a longer life cycle for the console. Long after Nintendo had released the Super Nintendo Entertainment System, games were still being released for the Famicom. Many of them were unlicensed, but gamers continued to buy them, keeping the console in living rooms well beyond Nintendo's expectations.

To exert control over the networks that Nintendo hoped would form around the NES, code was not a sufficient form of legislation. The patent office's granting of a patent provided one legal means by which force could be mobilized against those wishing to get around Nintendo's Seal of Quality. Nintendo was also careful in the copyrighting of the code that composed the 10NES chip. This provided them a second means of mobilizing the state to enforce compliance with the rules that they could not enforce themselves. Those wishing to pick the lock were now subject to litigation, and while it was still possible to reverse engineer the patented technologies, those doing so must be careful so that when they were sued, they could properly defend themselves in court against the state's enforcement of Nintendo's power.

While several companies did manage to circumvent the lockout capabilities of the NES, most did not. In Nintendo's most publicized legal loss, Galoob demonstrated that their Game Genie product for the NES made no use of copyrighted Nintendo technologies. The Game Genie merely altered the code being transmitted from cartridge to console. The Game Genie did not circumvent the 10NES lockout chip, but instead used the key device in the cartridge to allow normal booting of the NES. Galoob's technological and legal success, however, was unusual. More commonly, companies attempting to market games outside of Nintendo's new rule system paid dearly.[3] The most famous of these was a case involving both patent and copyright infringement, wherein a developer failed to jump Nintendo's legal hurdles.

Yet another case—Nintendo vs. Atari and Tengen—became the precedent for many of Nintendo's future legal claims. The case was foundational for all subsequent litigation to control the means of production even outside the game industry. The case is also particularly important because without it, the details of licensing arrangements mentioned in World 6-2 would have remained invisible. Unfortunately the case was exceptionally poorly played by Atari and Tengen, as is demonstrated in a brief excerpt from the court report. Needless to say, Atari went about "reverse engineering" the 10NES in a most harebrained[4] fashion.

Atari first attempted to analyze and replicate the NES security system in 1986. Atari could not break the 10NES program code by monitoring the communication between the master and slave chips. Atari next tried to break the code by analyzing the chips themselves. Atari analysts chemically peeled layers from the NES chips to allow microscopic examination of the object code. Nonetheless, Atari still could not decipher the code sufficiently to replicate the NES security system. . . .

In early 1988, Atari's attorney applied to the Copyright Office for a reproduction of the 10NES program. The application stated that Atari was a defendant in an infringement action and needed a copy of the program for that litigation. Atari falsely alleged that it was a present defendant in a case in the Northern District of California. Atari assured the "Library of Congress that the requested copy [would] be used only in connection with the specified litigation." In fact, no suit existed between the parties until December 1988, when Atari sued Nintendo for antitrust violations and unfair competition. Nintendo filed no infringement action against Atari until November 1989.

After obtaining the 10NES source code from the Copyright Office, Atari again tried to read the object code from peeled chips. Through microscopic examination, Atari's analysts transcribed the 10NES object code into a handwritten representation of zeros and ones. Atari used the information from the Copyright Office to correct errors in this transcription. The Copyright Office copy facilitated Atari's replication of the 10NES object code.

After deciphering the 10NES program, Atari developed its own program—the Rabbit program—to unlock the NES. Atari's Rabbit program generates signals indistinguishable from the 10NES program. The Rabbit uses a different microprocessor. The Rabbit chip, for instance, operates faster. Thus, to generate signals recognizable by the 10NES master chip, the Rabbit program must include pauses. Atari also programmed the Rabbit in a different language. Because Atari chose a different microprocessor and programming language, the line-by-line instructions of the 10NES and Rabbit programs vary. Nonetheless, as the district court found, the [F.2d 837] Rabbit program generates signals functionally indistinguishable from the 10NES program. The Rabbit gave Atari access to NES owners without Nintendo's strict license conditions. (Atari, Tengen, and Nintendo 1992)

This case begs the question, why was Atari trying to get around the limitation? Why not just talk to Nintendo? There is, of course, the possibility that Atari was simply bitter, having gone from the leader of the videogame industry to a player forced to work within the rules of another company. It is possible that Atari simply wanted to siphon off some of Nintendo's videogame profits. What neither answer gives us is any insight into the working realties in which Atari, Tengen, and Nintendo were operating. We have no way of knowing why companies were feeling compelled to work around Nintendo rather than with them.

One answer might be that the NES was so chock full of new and interesting technologies that developers were happy to buy into licensing schemes as a means of inserting themselves into the networks that were emerging—they played Nintendo's game to get a chance to play with new technologies. As with India's game studios, developers were willing to trade their rights for new and shiny technologies. One particular innovation that separated the NES from previous generations of consoles and might have inspired developers to work within the new Seal-of-Quality world of licensing restrictions was Nintendo's use of a new technology, a precursor to the now ubiquitous GPU. On the NES, Nintendo improved and simplified the way graphics were processed and delivered to the television screen with the Picture Processing Unit or PPU. This major design innovation also allowed the CPU of the console to spend more time doing game-related operations and less time handling graphics-related operations.

While it may seem that I am leveling blame for difficult work conditions solely at console manufacturers, I am not. Rather, I'm trying to elucidate that in the videogame industry, there is no boogeyman; closed and secret structures and networks are co-constructed. Game developers allow and perpetuate the current state of affairs. Put simply, game developers allow whatever game developers allow. People who create media, writing, and technology have long been concerned and interested in limitations placed on the distribution and use of their products. The methods of limitation in videogame have come to leverage the powers of the state more and more.[5] These limitations are also inherently limitations on the ability to produce for technologies covered under these methods. The more tightly distribution is controlled, the more tightly controlled production will be.[6]

At its simplest level, the cost of technology can do a great deal to ensure that users and distributors of videogames follow the rules set by videogame manufacturers. If you look at the technologies used in each console system and look for the cost of a given distribution medium it becomes apparent that when first released each was prohibitively expensive for an average consumer. In 1996, 64MB memory modules were not cheap, nor was it easy for consumers to create them or place them into plastic cartridges to fit into the Nintendo 64. This was one reason Nintendo opted to continue using cartridges in their systems like the one pictured in figure 6.4, despite Sony's move to use CD-ROMs in the competing PlayStation. In 2000 when the PS2 was introduced, the cost of a DVD burner was more than $4,000 and disks for those burners started at $40, nearly the cost of a videogame to

U.S. Patent Jan. 12, 1993 Sheet 2 of 10 **Re. 34,161**

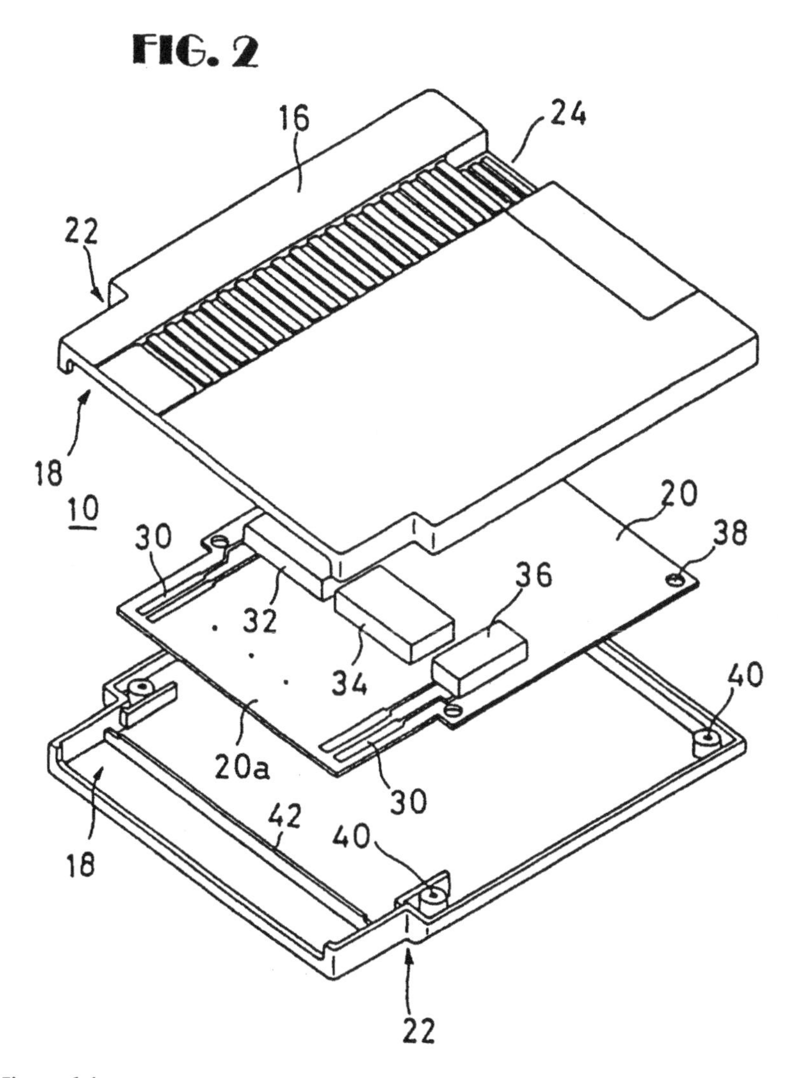

Figure 6.4
A schematic of a NES cartridge (Nakagawa and Yukawa 1987)

the average consumer. By the end of the PS2's life span however, DVD burners were a common add-on for new PCs. When the PS2 was first released, it was expensive or difficult to circumvent copy protection. By the end of the devices' life cycle, it was far easier and cost effective. The technology moves forward not just to provide new capabilities for game developers, but to also move on to platforms that are, for the most part, exclusive to content creators.

The same was true for nearly every medium on which console games were distributed. Of course this does not mean that either company did not also take precautions to reinforce the cost or difficulty of producing these items. Knowing full well that an all-out copyright assault would be expensive, drawn-out, and difficult, each company again employed the use of secretive and patented mechanisms to prevent production of unauthorized duplicates.

All of these technological and legal machinations have fundamentally altered the way games get developed. One specific way is that developers can't create these games on PCs without the same technologies that the consoles have. Thus, the NES also heralded the birth of a now-ubiquitous game development technology: the "DevKit."[7] DevKits were introduced so that game developers could create games for consoles where the hardware differed significantly from that of PCs. Nintendo developed technologies to bridge the gap between the PCs (where code was typically written) and the consoles (which ran the code). The complexity of these devices has increased dramatically with the complexity of consoles.

DevKits are also distributed with the software packages that simplify the process of game development. These range from SDKs that provide a set of software resources for developers to draw upon, to software tools that combine art, code, and data into a format that can be delivered and run on a DevKit. It is likely that the Nintendo Famicom, with accompanying disk drive, as pictured in figure 6.5, could well have served as an NES-era DevKit. DevKits can also include very basic technologies like compilers, IDEs, and debuggers, which are indispensable tools for game developers. Without these resources, the process of creating games can be much more complex. On a simple system like the NES or Game Boy Advance, hobbyists can overcome some of these limitations even though the resources typically available to developers are off limits. To be clear, DevKits are *leased* from the console manufacturer. Developers do not own DevKits. A lease can be revoked at any time and for a variety of reasons, and thus the very ability to develop games for these platforms can also be revoked. Fees paid to acquire a leased device are not returned.

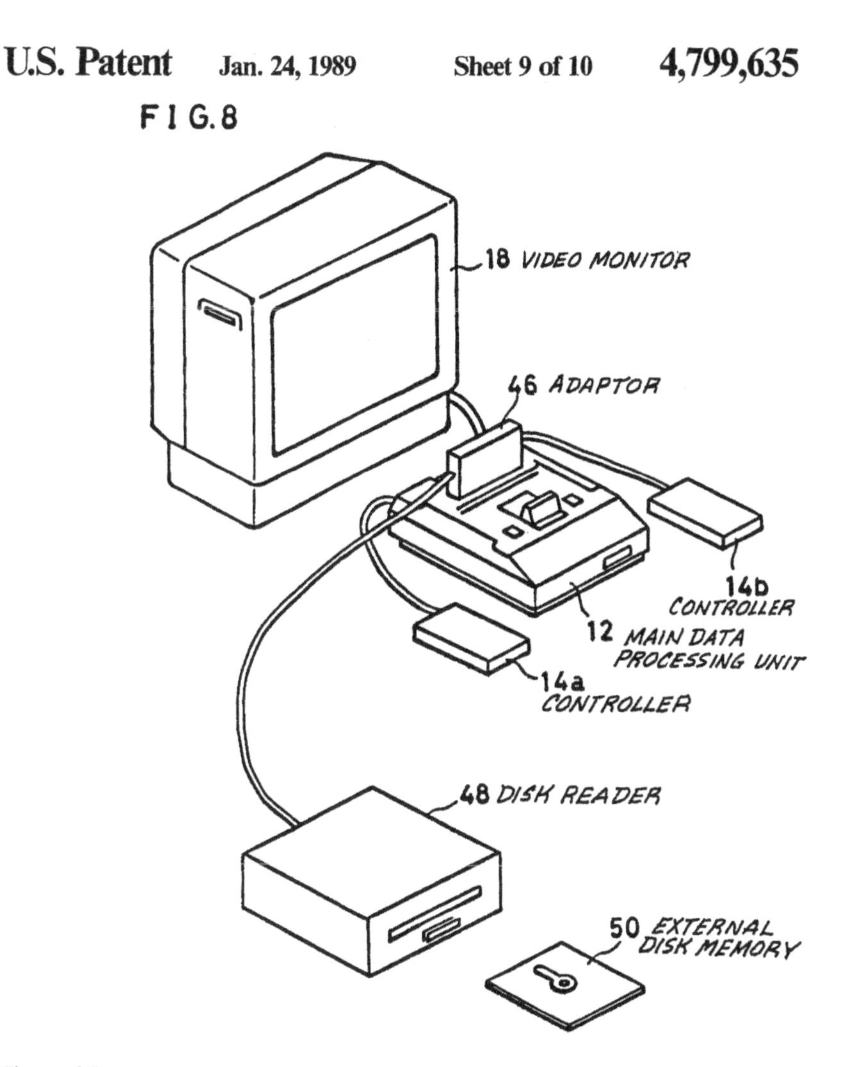

Figure 6.5
The Famicom combined with a disk drive: The NES DevKit

Accompanying the DevKits from manufacturers are NDAs that further limit game developers. These agreements legally prevent developers from distributing or sharing knowledge of or resources for these systems. While companies covered by the same NDA technically could share knowledge and resources, such openness is prohibited by the overarching industry emphasis on secrecy. This means that an unconnected producer cannot create a game for one of these devices; doing so requires the permission of the manufacturer. More on this in World 6-4.

Without DevKits, games can be developed and prototyped, but will later require massive changes to support the highly specific characteristics of a game console. Such an assertion reflects on distribution explicitly, but does not emphasize clearly that development necessitates these kinds of technologies. While distribution may be touted as "more accessible," this won't be true until DevKits are easily and cheaply acquired, or SDKs no longer depend on them. Game studies scholars and independent game developers have all lamented the limits this places on developers.

> Both physical and digital distribution rely on independent developers' ability to make games for the new platform in the first place. While Nintendo has been quite vocal about its intention to support independent developers, including offering Wii dev kits for under US $2,000, Nintendo of America has also said that it won't start reviewing independent developer applications until January 2007—which means that only those developers with publishing contracts or special invitations actually have them. (Bogost 2006)

Even by the summer of 2007, independent developers were unable to acquire these devices, despite announcements that indicated otherwise. The rhetoric used in this announcement harkens to the one used by Microsoft with the announcement of their XNA Express endeavors mentioned in World 3's Foss Fight. Under close examination, the reality of console technologies locking out developers indicates that very little is actually changing.

And this failure to broaden access really means that Nintendo's announcement was made to indicate an intention to distribute independent and original games on their online network, much like Microsoft and Sony had already been doing on their networks. So while Microsoft's XNA technologies are actually creating opportunities for developers to share resources and technologies, Nintendo's approach to production control remains the same, despite the similarities in the rhetorical framing of their press releases. The mechanisms for controlling production continue.

If the ability to gain access to the game development technologies for these systems has been difficult for developers in established industries, it is even more complex in countries like India with emerging industries. The legal, technological, and political maneuverings unflatten the global playing field. While I was in India and when I left at the beginning of 2007, there were precisely two Nintendo DS DevKits in all of India, both at the same company, which had recently been acquired by a rapidly globalizing US-based publishing company. The studio's engineers and artists were trying to get their heads around this more limited technology with SDKs they had never seen before. It was one of those odd moments where my past developing videogames suddenly became useful during fieldwork. Having worked with Nintendo SDKs during the time of the N64, I recognized some of the techniques and standards that were being used. For the developers in India these were new concepts, unexplained and undocumented (except in cursory terms and in ways not useful for developers making games). When asking questions on the private forums used for developer discussions, they frequently encountered hostile responses like, "How can you not know this?" So I spent some of my time working with engineers offering what aged and blurry knowledge I did have to assist in their development work. In some cases I sat with engineers while they wrote code. Other times we talked at whiteboards. Sometimes I asked artists how they were generating their art and getting it into the game. How did they play the game? Their absolute lack of the knowledge held by other Nintendo developers yet but not included with the two Indian DevKits precisely illustrates the point that industry structures control production and prevent any circulation of knowledge regarding what it takes to create games for console systems.

In many respects, in a platform studies sense, the NES facilitated certain types of social and technical support systems that surrounded it. Thus, in addition to the ways "hardware and software platforms influence, facilitate, or constrain . . . computational expression" (Montfort and Bogost 2009, 3) they reach further. Platforms matter not just for the technical creativity of artists, programmers and designers, but also for their broader social context. It is the broader social, political and economic context that I would argue that platform studies often misses, by focusing so tightly on the technology of the devices examined. As such, it is that setting that provides a kind of historicity for a device that also shaped its development. For example, in the platform studies examination of the Nintendo Wii (Jones and Thiruvathukal 2012), no mention is made of the well known 2007 GDC rant delivered by Chris Hecker, where he characterized the device as "two [Nintendo] Game Cubes duct-taped together." This statement nearly cost him his job, despite the goal of the GDC rants to serve as "off the record" moments. The idea that the Wii was more like the Game Cube than it was different was particularly humorous to those in the audience working on the Wii. Early DevKits for the Wii were actually Game Cube DevKits with the addition of periphery hardware and updated SDKs. To those developers with that knowledge, the Wii really had started its life as a Game Cube. Thus, the Wii, as a platform remains inherently linked to the Game Cube and its shaping of that platform. Another such example is that the "waggle" recognition software distributed by Nintendo for developer, was not actually included in early DevKits, but came later in the life of the console as developers identified the need.

Secrecy and protected power networks also have the secondary effect of encouraging Indian game development companies to focus on the production of art assets. Because the tools of game art production, *Max* and *Maya*, are relatively standard, developers in India are already familiar with these tools and are able to produce artwork for United States and Western European companies. However, because the tools that would allow engineers to create code for other systems and begin generating tools for designers to bring these aspects together are unavailable, many companies are forced to focus on those aspects most amenable to offshore outsourcing. This meant that rather than being able to bootstrap themselves into the global game industry, they are structurally positioned to act as art production houses. Because of this, as mentioned in World 5, many of the Indian game studios shift their focus to the production of game titles for mobile cell phone devices, currently the smallest market within the broader game industry. Of course this is changing, as the introduction of *Android* and iOS based phones and their accompanying digital distribution platforms has caused a major shift for both content creators and users.

Many of the other technologies that compromise consoles and PCs have actually converged, and the lockout chip, 10NES or otherwise, perseveres. It is the main limitation for developers hoping to distribute their games on consoles.[8] It also prevents new publishers from challenging existing ones; and even when they do, the standard approach is consolidation and acquisition. The distribution of DevKits fits nicely into the networked structure throughout the videogame industry. The big players lease these DevKits for large sums of money and use them or distribute them to developers making the games for consoles. Even consoles purported to use "standard CD-ROM" drives (Malliet and Zimmerman 2005) place limitations on developers.

Regardless of how a device characterizes its effect, if a technology being touted as a controller of distribution or production, the net effect is actually control of both. These technologies and their connection with political and legal structures disable the ability of producers to use, learn from, and share experiences with one another. Such technologies reinforce the prejudice of developers that the work of game development is somehow different and separate and encourages the maintenance of secret societies throughout the industry. And while corporations scream that the market must be allowed to function without regulation, they simultaneously leverage government regulation to alter the market specifically in their favor and at the expense of those they depend upon for production. Only those who have been approved and are inside the networks should be allowed a distributed voice: this is an inherently anti-market approach. However, corporations cannot enforce these systems without the intervention of the state.

World 6-4: The Non-Disclosure Agreement (NDA)

The non-disclosure agreement is a peculiar animal in the videogame industry. It is the embodiment of secret keeping and, at the same time, it is one of the most frequently breached documents, often for the benefit of game developers. The informal conversations that developers have with one another form the foundation of institutional memory and professionalization. Yet in these undocumented intrastudio discussions, the NDA still carries significant power.

There has been significant recent backlash to the NDA as videogame console manufacturers have attempted to make their hardware available to university programs, so that budding engineers can play instrumentally with the hardware available prior to entering the game industry. At the same time, these devices have remained under NDA, which means that for the most part students cannot post any source code for the projects developed under the program (Danks 2008; Orland 2008). Students can share information with one another, but cannot put their findings into conversation with the broader communities of hobbyist and independent game development.

NDAs pervade the game industry. They exist between organizations and individuals and among organizations. What results is an industry founded on and bounded by silence, in part based on a desire for secrecy (as mentioned in World 2) and in part based on necessity. Developers talk mostly in generalities, so deeply have they internalized the code of not disclosing. Even when companies ultimately belong to the same parent company, they continue to not share information based on accustomed practice. Developers acquire a built-in paranoia about what can or cannot be discussed, resulting in a kind of constant self-policing that resides at the core of our consent to hegemony. The fear that another game developer, who is hard at work on their own project, might become aware of the project you are working on operates as an effective silencer within an industry where a competitor might incorporate your super-secret, earth-shattering technology to gain more sales than you. Publishers want to control information flows to ensure their marketing departments manage the public relations of a soon-to-be-released game. Hardware manufacturers fight to ensure that information about the underlying technologies of their devices are not leaked to other manufacturers or hackers hoping to break the copy protection on their devices and extend their functionality beyond that of their intended design (Androvich 2007). Within the industry, fear, or at least self-censorship operates at all levels of development to help manufacturers, publishers, and studios maintain secrecy.

At the same time, game developers are often desperate to find more information about how others develop games, and the Game Developers Conference (GDC) is one event where information flows more freely, though still consciously constrained by the legal implications of NDAs. At one such event I asked a group of game developers who were all working on the Nintendo DS, "How many of you have written an XML parser for the DS?" Nearly every engineer in the room raised their hand. Discussion between developers ensued and at least two open source solutions, expat and TinyXML, were found to have been ported numerous times. One must ask how many of the technological and engineering problems found in game development have already been solved broadly by the Free/Open Source Software movement (F/OSS).[9]

Thus, the NDA embodies a foundational memory-loss system. While organizations ought to protect their intellectual property, often the uniqueness of a given technology is much less than it is often given credit for. In only a few, truly unique, instances have developers opened up and begun talking about the technologies that form the foundation of their game development practice. One example of a game studio bucking this trend is Insomniac Games, which has started its "Nocturnal Initiative™," open sourcing numerous aspects of their videogame development process and the tools that form the connective tissue between their numerous systems. Insomniac is a highly successful game studio, with numerous highly successful game titles under their belt, including *Ratchet and Clank*, *Resistance*, and *Spyro*. Their efforts with Nocturnal Initiative™ are aimed at unveiling the tools that might save other developers time: "Nocturnal Initiative™ is not a game engine. The libraries provided here are potentially very useful for developing a game engine, but we want to avoid the 'all things to all people' that so often results in overly complex and/or under-performing monolithic engines. Instead, we want to provide a useful toolbox for the professional game developer" (Insomniac Games' website quoted in Evans 2008). It is precisely this kind of commitment to break the culture of secrecy perpetuated by the pervasive NDA traditions that stands to transform game development practice if adopted more broadly.

Having observed the effects of a videogame industry that has labored under a culture of NDA for nearly thirty years (since the introduction of the NES and licensing), it is surprising that when more typical software developers encounter similar limitations on new devices, they react very differently from game developers. After the release of the iPhone 3G in 2008, which added the ability to create custom applications for Apple's App Store, developers quickly reacted strongly to being unable to share information with one another about how to best work with the device (Chen 2008). In fact, the website fuckingnda.com was created within two months of the iPhone 3G's release and offers a place in which software developers can rant about their limitations. Strangely, game developers have worked under similar restrictions for nearly three decades and have not balked at their lack of ability to share practices with one another. NDA restrictions are constantly referenced at conferences or during informal conversations among developers, who nevertheless continue to work under the yoke of those systems. The culture of secrecy and the culture of the NDA has pervaded development so fully that developers no longer question the logic of the limitations or what their implications might be.

World 6 Boss Fight: Institutional Alzheimer's

The game industry needs to become more open for its own survival and growth. Ultimately this openness must occur both at the lowest and highest levels. Game developers must be able to converse broadly about the practice of game development. Publishers and manufacturers need to be able to differentiate between talking about how one goes about making games and "giving away" a game. Many software companies have made numerous aspects of their work and work processes available online to foster a community of practice. The important difference is that for game companies this openness would go beyond releasing the "source code" of a game. It would also document and reveal how artists and designers went about creating and working within the source code of a game: how they created content and data, which then resulted in a game. Useful discussions should involve samples of real data that artists worked on and their process to get it into the engine. Designers should be able to document and explain how data combined with artistic assets and how it mobilized the source code to create a game.

Much of this information already exists, in studio and corporate wiki sites, and could easily be shared. The argument that developers simply do not have enough time to do the work of documenting their processes and structures is simply not true. They already do it internally. The real hesitation lies in a fear of losing competitive advantage. Yet sharing the information more broadly can only make developers more effective at the practice of game development. Portions of their internal wikis can be released more broadly, perhaps delayed until after the release of a game title, for huge gains in the industry's knowledge base and capabilities. Developers could be spared reinventing the wheel and could use their talents to push the industry further.

Further, these wikis can serve as the foundation for fostering new developers interested in working with those practices. Young developers interested in becoming part of these game studios can become the intermediaries between the company and its community. Aspiring game developers can uses these insights in their own quests to create videogames. Rather than making the same day-to-day mistakes or misunderstanding how the process functions, they can learn from some of the lessons of the nearly thirty years of videogame development history. Rather than being seen simply as a training or proving ground for new unpaid talent, wikis can be used as a space for developing collaborative skills among new paid developers who then more quickly and easily learn how to use the tools and work with others both in and outside the company.

Sharing portions of existing developer wikis will also encourage a broader understanding of what goes into making games. Artists, designers, engineers, and managers who already participate in the generation of these resources, will make the work of game development more visible, and in so doing can make explicit the cross-disciplinarity of the endeavor. The importance of process, tools, communication, and collaboration can be clarified and taken to new levels. The imagination of game developer as computer scientist will no longer reign supreme, replaced, instead, by a new generation of developers coming to the industry with diverse backgrounds, educations, and talents.

Once networks and structures shed the veil of secrecy, teams will have the opportunity to make the numerous design decisions and their impacts visible. Making more transparent the effects of sudden shifts of scope or design dictates from other interests can provide insight into the lived realities of game development. Collectively, this information may encourage developers to work with particular manufacturers and publishers in favor of others that detrimentally affect the work practices of developers.

Transparency may also help publishers and manufacturers understand why developers are resistant to dictated shifts or changes. Improved visibility could provide publishers and manufacturers with insight into when and why studios or development teams are not moving forward successfully. Transparency cuts in numerous directions, all of which would seemingly benefit the game industry.

Transparency will begin to demystify the game development process, so new conversations can begin about these processes, discussions that are explicit and clear rather than general and vague. Companies can discuss aspects of game development that have historically remained closed. More than anything, opening up will encourage game developers to think of themselves in a broader collective context, rather than as individuals in individual studios scraping against all odds against their fellow developers.

The unlocking of the videogame industry would create new opportunities for creativity and entrepreneurship from new locations around the world. If US-based game development companies are willing to work with globally located companies for their artistic production, then these companies should be provided the access necessary for further engagement, and the credit necessary to build their prestige.

Even veterans of the videogame game industry, like the vice president of operations and development for Big Huge Games, which created *Rise of Nations* for the PC, a project that took more than three years and nearly eighteen hundred files to create, express their frustrations over the lack of institutional memory. As Train notes, what irks him most is as follows: "Not listening to all the other 'Postmortems' ever printed in *Game Developer*. The 'Postmortems' are the most widely read feature in *Game Developer* around Big Huge, and yet somehow we still managed to make many of the mistakes developers are cautioned against in these pages" (Train 2003, 40).

Silence, while preventing the disclosure of outside influences, encourages developers to discount the experiences of others as distinctly unique from their own experiences. Tim Train's comments indicate that while postmortems may not be directly applicable or importable into each and every studio, there is something to be learned from those experiences in the context of any game development company. This also requires that developers build in the time to address the concerns of previous projects and perhaps even incentivize sharing practices during development schedules.

Secrecy clearly hinders the passing on of information to those interested in joining the videogame industry, be they domestic or international game developers. Because common practices are not documented or circulated broadly, students, hobbyists, and independent developers are left to reinvent the wheel and relearn practices that ought to be commonplace in game development practice. The Lead Designer of the game *Asheron's Call*, which took four years to create, comments on the lack of experience and communication difficulties that arise between disciplinary groups in the workplace.

> Many of the employees were students immediately out of college, or even college students completing a work-study program. . . . It was nearly impossible for team leads to give realistic schedule estimates for tasks, since few of us had experience in professional software development. It was also initially difficult to get different teams from the programming, art, and design departments to communicate regularly with each other. (Ragaini 2003, 307)

As sociological studies of science indicate, estimation in particular requires experience (Pinch, Collins, and Carbone 1997). If developers made it clear that estimating was something that a developer should be thinking about when approaching tasks, it would be more transparent. An educational website could note, "Before you begin this task, estimate how long you think it will take you to complete the exercise." If estimates are routinely ignored because they are either dramatically under- or overestimated, how does that affect project deadlines? New conversations about practice can begin, ideally resulting in more realistic timelines and less crunch. But this can only happen if there is real education of new developers and if there is genuine effort at understanding and fixing current practices, all of which hinges on the end of industry-wide silence.

The (in)ability for game developers to learn and share information about game production practices severely limits the capacity for the industry to mature. This is largely due to the restrictive practices surrounding the tools necessary for the production of videogames. NDAs prevent developers from sharing source code, tools, or information about how to navigate and apply these devices. These NDAs are disguised under the category of "licensing" at the level of the publisher and console manufacturer. Ultimately it is the manufacturer who demands that information about the console not be shared more broadly, though these agreements can also be passed down from manufacturer to publisher, and then to developers.

The time for broader participation is at hand. For too long, game developers have seen their culture as largely Western or Japanese. Web 2.0 as broadly presented is an indicator of this changing relationship. While I may quibble with the particularities of what Web 2.0 is precisely, at its core it is a changing relationship between "user" and "producer." There is a fundamental difference, but that difference is dependent upon ideas that the game industry has not yet embraced: access, standards, openness, participation, and remixability. The new technological paradigm demands core modes of operation that are still anathema to game development, and this needs to change.

Sony's Phil Harrison continues to speak about "Game 3.0" as the videogame industry's version of Web 2.0, yet the industry's reality is something quite different. It remains vetted to a broadcast model of game design, development, and distribution, which is an inherently outmoded, previous generation system. "Community" and "customization" remain limited to the rather small sandboxes provided for players. This is quite different from Web 2.0, where companies may not always be happy at what their users produce or reproduce. If the game industry currently allows for very small, highly controlled and lifeguarded sandboxes, to truly embrace the possibilities of Web 2.0, the game industry needs to think more in terms of a vast desert complete with wastelands, but also an oasis that flourishes outside the standard protocols and rules that dominate the game industry.

Despite the slide delivered by Sony's Phil Harrison at GDC 2007 containing the words, "open, extendible, customizable, collaborative, audience-driven, localized," the reality is that the only openness that has been realized in the four years since his address's delivery is perhaps "content

Figure 6.6
Sony's Phil Harrison speaks of "Game 3.0" at GDC 2007

creation" and "commerce," and even those remain significantly limited by the networks of connection to Sony (figure 6.6). Web 2.0 depends on a backbone of open technologies, formal standards, and a community of developers whose actions reflect the terms, not mimic them. Even post-dot-com-bust, the World Wide Web remains the realm of innovation, more so than the game industry center.

Harrison's slide was an appeal to the draw of Web 2.0 that attempted to capture the excitement being generated by a similar, but very different image. Perhaps unsurprisingly, Phil Harrison subsequently left Sony Computer Entertainment for Atari and has labored to encourage broader acceptance of open technologies among game developers. Harrison spoke at the Unity game engine's Unite conference in 2008, discussing the importance of open and accessible game technologies.

Numerous words appear in figure 6.7 that do not appear in Sony's vision of Game 3.0. Most important are the words, "standardization, open APIs, data driven, design, and participation." These words do not appear in Sony's understanding of what the future holds. Regardless of the "reality" of Web 2.0, something is happening online. YouTube, Google, and Really Simple

Figure 6.7
Web 2.0 graphic released under the creative commons copyright

Syndication (RSS) refocus media on the user as co-conspirator in the design and development process. The secret society of "production" is always already in doubt in this space, yet its long term viability is dependent upon openness, standards, and remixability. This is the everyday life that game developers and the game industry must embrace for sustainability.

Developers face structural barriers to openness. These barriers will require significant changes by console manufacturers and publishing companies, who must switch the power network that has been in place since the release of the NES. These companies technologically and legally barricade access to distribution channels. Yet the fears that led Nintendo to close off the NES no longer threaten it. It is time for the game industry to embrace a culture of openness and sharing that facilitates the growth of the medium, rather than its continued stagnation. Manufacturing companies have largely been playing the same game for the last thirty years and they will likely be hesitant to play by new rules. In a more open production space, there will be greater competition and the possibility for content that manufacturers do not endorse or approve. However, this is already the industry's reality. The only differences are that game developers will be more able to share knowledge and resources and users will have a new opportunity to embrace and explore a medium they already care so deeply for.

Console manufacturers and publishers will likely claim that opening up routes of production will increase piracy. However, this does not seem to correspond with Microsoft's experiences on the Xbox 360 with the XNA Toolkit. Many of the efforts by other console manufacturers to stem the tide of piracy have had the shadow-effect of silencing new lines of independent production. When not arresting or shutting down companies that make these technologies possible, manufacturers stem the tide of broader involvement by blocking homebrew developers with "updates." Upgraded features for existing devices tempt users to install updated "firmware" (software upgradeable hardware) technologies that render devices with homebrew software unusable or no longer accessible. Despite this, time and again, users are able to bypass new mechanisms, again making them open to user modifications, new capabilities and in some cases, piracy. It seems illogical to invest so much time, energy, money, and legal resources into preventing the modification, hacking and repurposing by users who will continually work to make their consoles function how they desire. These developers are bringing new functionality to these devices at no cost to the manufacturer. In the end, the risk is that content that has not been approved or paid licensing fees will make its way to these devices. However, the continued control of development hardware or DevKits will likely encourage most developers to pay licensing fees and retain their connections with manufacturers as the risks associated with going rogue are often too high. This will simultaneously allow indie developers to further the industry, rather than thwart it. With open avenues of independent production, the risk for working outside the boundaries of the industry will prove much higher and as a result, these new developers will further the industry rather than attempt to thwart it.

Though developers and companies within the videogame industry will both have to give up certain elements of control and positions of power that the existing structure encourages, the benefits outweigh the drawbacks of previous approaches. Now, more than ever, the game industry must come to understand that openness will benefit the sustainability and maturity of the game industry in the long term. To continue down a path lacking standards, emphasizing secrecy, and subject to aging patent and copyright law that largely does not understand new interactive media, is to progress down a path where repeated and compounding errors and risks will continue to eat away at the game industry. Refusing to change disrespects the creative work poured into making the place that videogames have come to occupy in broader global culture.

The boss fight for this level is ultimately the one that game developers and the game industry may be the least willing to play, but it might be the most beneficial fight to experience. Creating forums for broader cooperation across the videogame industry will create opportunities for learning and sharing. There is the potential for transgressive possibilities, which may be viewed as undesirable. This is an inevitable consequence of encouraging more participation in the creation of videogames rather than a closed off world of limited participation. Ultimately however, existing structures enable corporations to maintain control over their "official" networks.

How would PC and mobile sectors, those which have had the most penetration by global players outside the typical networks of access, benefit from these efforts at documented and shared networks and systems, while the console game industry game would not? In part, because of the familiar dismissal of a single platform:

> The idea of a single platform for the videogame console industry has been kicked around nearly every videogame cycle. Publishers would gain leverage over console manufacturers or forego licensing payments altogether with a collaborative organization to develop standard hardware and software specs and requirements. I don't believe such a consortium could bring about a single console system. Business models and publisher strategies are too divergent to enable agreement on a hardware platform. (Staff 2007)

Yet openness could be a boon to game developers if they were to demand a common open reference platform rather than a single console device. This boss fight is an argument for information sharing, not total standardization. The continued labeling of the console game industry as different or unique forces the argument to fold back onto the work practice of game developers. They must come to see their lives as intimately connected rather than outside of traditional forms of work and social-technical practice if the industry is to grow.

World 7: Disciplining the Industry's Actor-Networks

Box 7.1

```
#: SET DEMO_MODE 1
AUTHOR_DEV_DILEMMA: [REDACTED]
```

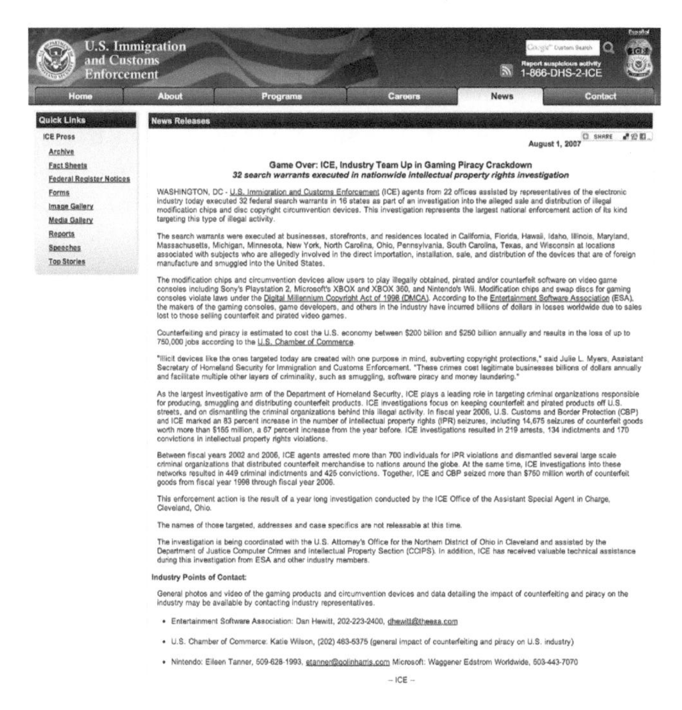

U.S. Immigration and Customs Enforcement

Report suspicious activity 1-866-DHS-2-ICE

Home | About | Programs | Careers | News | Contact

Quick Links

ICE Press

- Archive
- Fact Sheets
- Federal Register Notices
- Forms
- Image Gallery
- Media Gallery
- Reports
- Speeches
- Top Stories

News Releases

August 1, 2007

Game Over: ICE, Industry Team Up in Gaming Piracy Crackdown

32 search warrants executed in nationwide intellectual property rights investigation

WASHINGTON, DC - U.S. Immigration and Customs Enforcement (ICE) agents from 22 offices assisted by representatives of the electronic industry today executed 32 federal search warrants in 16 states as part of an investigation into the alleged sale and distribution of illegal modification chips and disc copyright circumvention devices. This investigation represents the largest national enforcement action of its kind targeting this type of illegal activity.

The search warrants were executed at businesses, storefronts, and residences located in California, Florida, Hawaii, Idaho, Illinois, Maryland, Massachusetts, Michigan, Minnesota, New York, North Carolina, Ohio, Pennsylvania, South Carolina, Texas, and Wisconsin at locations associated with subjects who are allegedly involved in the direct importation, installation, sale, and distribution of the devices that are of foreign manufacture and smuggled into the United States.

The modification chips and circumvention devices allow users to play illegally obtained, pirated and/or counterfeit software on video game consoles including Sony's Playstation 2, Microsoft's XBOX and XBOX 360, and Nintendo's Wii. Modification chips and swap discs for gaming consoles violate laws under the Digital Millennium Copyright Act of 1998 (DMCA). According to the Entertainment Software Association (ESA), the makers of the gaming consoles, game developers, and others in the industry have incurred billions of dollars in losses worldwide due to sales lost to those selling counterfeit and pirated video games.

Counterfeiting and piracy is estimated to cost the U.S. economy between $200 billion and $250 billion annually and results in the loss of up to 750,000 jobs according to the U.S. Chamber of Commerce.

"Illicit devices like the ones targeted today are created with one purpose in mind, subverting copyright protections," said Julie L. Myers, Assistant Secretary of Homeland Security for Immigration and Customs Enforcement. "These crimes cost legitimate businesses billions of dollars annually and facilitate multiple other layers of criminality, such as smuggling, software piracy and money laundering."

As the largest investigative arm of the Department of Homeland Security, ICE plays a leading role in targeting criminal organizations responsible for producing, smuggling and distributing counterfeit products. ICE investigations focus on keeping counterfeit and pirated products off U.S. streets, and on dismantling the criminal organizations behind this illegal activity. In fiscal year 2006, U.S. Customs and Border Protection (CBP) and ICE marked an 83 percent increase in the number of intellectual property rights (IPR) seizures, including 14,675 seizures of counterfeit goods worth more than $155 million, a 67 percent increase from the year before. ICE investigations resulted in 219 arrests, 134 indictments and 170 convictions in intellectual property rights violations.

Between fiscal years 2002 and 2006, ICE agents arrested more than 700 individuals for IPR violations and dismantled several large scale criminal organizations that distributed counterfeit merchandise to nations around the globe. At the same time, ICE investigations into these networks resulted in 449 criminal indictments and 425 convictions. Together, ICE and CBP seized more than $750 million worth of counterfeit goods from fiscal year 1998 through fiscal year 2006.

This enforcement action is the result of a year long investigation conducted by the ICE Office of the Assistant Special Agent in Charge, Cleveland, Ohio.

The names of those targeted, addresses and case specifics are not releasable at this time.

The investigation is being coordinated with the U.S. Attorney's Office for the Northern District of Ohio in Cleveland and assisted by the Department of Justice Computer Crimes and Intellectual Property Section (CCIPS). In addition, ICE has received valuable technical assistance during this investigation from ESA and other industry members.

Industry Points of Contact:

General photos and video of the gaming products and circumvention devices and data detailing the impact of counterfeiting and piracy on the industry may be available by contacting industry representatives.

- Entertainment Software Association: Dan Hewitt, 202-223-2400, dhewitt@theesa.com
- U.S. Chamber of Commerce: Katie Wilson, (202) 463-5375 (general impact of counterfeiting and piracy on U.S. industry)
- Nintendo: Eileen Tanner, 509-628-1993, etanner@golinharris.com Microsoft: Waggener Edstrom Worldwide, 503-443-7070

– ICE –

Figure 7.1
Screenshot of US immigration and customs enforcement website (ICE 2007)

```
AUTHOR_DEV_DILEMMA: [REDACTED]
#: SET DEMO_MODE 0
```

World 7-1: Software Is Society Made Malleable

In the collected volume *A Sociology of Monsters*, Bruno Latour authored an essay titled "Technology Is Society Made Durable" (Latour 1991). The argument that technology encodes and enacts ideas about how society ought to function was an important one. Yet technology has changed since 1991, in ways driven by ideas introduced by the game industry, and in ways we must examine. I believe scholars of science and technology have missed something critically important with regard to the software and hardware of technologies, in part because the two are now so interwoven. Updatable firmware, flash memory, solid state disk drives, network connectivity, and updatability—all of these belay a significant change underlying many technologies. This challenges in a fundamental way how technologies have begun to enter into dialogues with users. A device's capacities can expand and contract without the consent of the user as new software/firmware updates are applied. What was possible at one moment may become disabled without a user's consent. Functionality may be added to encourage new modes of interaction. Manufacturers, publishers, and authorized developers can, if they so desire, change how a device or software functions, and if the user is unhappy about those changes, leave them with no alternatives. In short, technologies have become more subject to change over time, based not on how users want to use devices, but based on how companies *expect* users to employ those objects. This contrasts with the devices scrutinized by Latour, which did not have this capacity, and thus many devices are put to use in ways not imagined by their creators.

Latour grapples with notions of "power" and "domination" through the lens of actor/actant-network theory in the context of socio-technical assemblages. At its simplest, his argument is that technology and society are intimately connected. Technology does not "impact" society, rather it is constituted in ways that are socially situated and society takes those technologies up in ways that are nondeterministic. Technology and society are intimately connected. Historically, this was both a theoretical and methodological turn toward understanding technology in the making.

> Unlike scholars who treat power and domination with special tools, we do not have to start from stable actors, from stable statements, from a stable repertoire of beliefs and interests, nor even from a stable observer. And still, we regain the durability of social assemblage, but it is shared with the non-humans thus mobilised. When actors and points of view are aligned, then we enter a stable definition of society that looks like dominations. When actors are unstable and the observers' points

> of view shift endlessly we are entering a highly unstable and negotiated situation in which domination is not yet exerted. . . . It is as if we might call technology the moment when social assemblages gain stability by aligning actors and observers. (Latour 1991, 129)

When Latour says "dominations" in this context, what he is really talking about is stable structures of power. This focus on technology as a point in time of stability is perhaps at best evocative of a Gramscian kind of hegemony. However, I think this is a weak use of the term domination, because when we're talking about technologies and the maintenance of actor-networks, some domination is more dominating than others.

As previous worlds have made clear, I have studied game development, game developers, and the game industry for years and my fieldwork has spanned from legitimate corporate game developers at studios owned by Activision to far flung field-sites in India (some owned by Activision and others owned by independent Indian developers). I have also looked closely at small homebrew and independent game developers in the United States, as my informants indicated that it was an important community where new developments were occurring. What is interesting is that domination is experienced in these more peripheral communities in a much stronger sense than in more conventional game development companies. This isn't to say that the same structures and dominations don't affect rank-and-file game developers, just in different ways. Before I launch into a discussion of homebrew developers in World 7-4, I want to be clear that while such enterprises might seem disconnected from the game industry proper, this is not the case. Homebrew developers experience the disciplining effects of structure that affect the whole industry, but feel the affects more acutely. Homebrew developers exist at the outer edge of the swerve that software and code creates for us as social analysts, which means domination that encourages stability in the center of the industry throws homebrew developers wildly about on the fringes of game production.

Throughout this world, you'll notice that I use the chiasma of software/code and silicon/hardware. I do this deliberately, because it is important to remember that silicon and electrons are part of this equation as well, especially since my informants (as quoted in earlier worlds) note that they know the very underlying structure of the hardware. As Mike Fortun puts it, "Science is made, but it is not made up" (Fortun 2007, 147–148). The same is true in game development as well. Software/specifications/etc. are made, but they are not made up. We cannot forget that though code is constructed, it is also linked to electrons, silicon, and other elements of the physical world.

At this intersection, then, of constructions and their underlying realities, we need ways in which to understand the system of structures that frame the work of developers in the videogame industry. This approach is heavily rooted in what I tend to think of as John Law's "heterogenous engineering" version of Actor-Network Theory. This framework provides an initial set of questions that enables us to start making sense of the numerous aspects of the game industry that come together in what often appears to be a jumbled mess. Put simply, theory serves as a primer for the parsing of this data. In particular, for a way to parse the systems that define the game industry we can adapt Law's attention to three particular aspects of a technological system: "An approach to these questions that stresses (1) the heterogeneity of the elements involved in technological problem solving, (2) the complexity and contingency of the ways in which these elements interrelate, and (3) the way in which solutions are forged in situations of conflict" (Law 1989, 111).

Heterogeneity? Complexity? Conflict? Domination? This is the stuff game technology is made of, and this world begins with an argument that software/code is part and parcel with what Omi and Winant might call "projects" or "formation processes [as] occurring through a linkage between structure and representation" (Omi and Winant 1994, 56). You can also think of the software/code creation as a kind of Pickerian "mangle" or Deleuzian "assemblage," if that seems more your cup of tea, given the observations in previous chapters on the preproduction and production shows. But, as we allow various terms to frame software/code, note that some actor-networks are more capable of mobilizing the abstract war machine, or as Wendy Brown may call it, "The Prerogative Dimension of the State . . . The [s]tates's 'legitimate' arbitrary aspect: extralegal, adventurous, and violent" (Brown 1995, 186) component. This hearkens back toward those observations in World 6, examining NDAs, lockout chips and court cases. These systems are intimately linked together.

Ultimately, it is this contradiction, that, at least in the context of the videogame industry, the systems are not always passages from the microscopic to the macroscopic, or emergent. As we saw in the Nintendo/Atari case, force can be exerted from numerous directions. Developers understand their craft as critically linked with practice, the NDAs of World 6, and the culture of secrecy that encourages developers to hold tight to their secrets. The pace of development that prevents real reflection defies a singularly micro-to-macro or macro-to-micro mapping. This leads me to question Latour's assertion:

> The narrative should also account for another little mystery: the progressive passage from the microscopic to the macroscopic. Network analysis and field work have been criticized for giving interesting demonstrations of local contingencies without being able to take into account the "social structures" which influence the course of local history. Yet, as Hughes has shown in a remarkable study of electrical networks (Hughes 1999, 50–63) the macrostructure of society is made of the same stuff as the micro-structure—especially in the case of innovations which originate in a garage and end up in a world that includes all garages—or, conversely, in the case of technological systems which begin as a whole world and end up on a dump. The scale change from micro to macro and from macro to micro is exactly what we should be able to document. (Latour 1991, 118)

Sometimes, power *is* administered from the top down. We *can* see top-down domination occurring in the game industry, and to chalk it up to "the same stuff" neglects the power found in these networks. Such a cycle feels more deliberate when we're talking about cross-nation, simultaneous, coordinated raids on corporations and individuals because there is a macrostructure here. It isn't just bottom-up; there is top-down happening. But since there is no clear procession from the micro to the macro, and since the videogame industry is much more mangled, I prefer to think of the way power is expressed in the game industry in terms of Dorothy Smith's notion of ruling relations. As she notes: "The intersection of everyday local settings and the abstracted, extra-local ruling relations is mediated by the materiality of printed and electronic texts. The texts integral to the social organization of these forms are complemented by technologies or disciplined practices that produce standardized local states of affairs or events corresponding to the standardized texts" (Smith 1999, 73).

Smith's framework acknowledges that local affairs may be quite different from the texts of the extra-local, yet that does not negate the force of those extra-local texts. For example, VV's interest in developing internal tools that spanned projects and platforms is not the same as other studios that purchase and adapt tool-chains or those that develop new tools and pipelines for each project. The local is always structured in some way by those extra-local texts, often in ways that are perceived as having derived from the local. The extra-local ruling relations are mediated and experienced. There is a disconnect between them, true, yet one cannot simply say that the extra-local is a product of the emergent local situations. As Smith elaborates: "Ruling relations in general, are, ontologically, fields of socially organized activity. People enter and participate in them, reading/watching/operating/writing/drawing texts; they are at work, and their work is regulated textually; whatever form of agency is accessible to them is accessible textually as course of action in a text-mediated mode" (Smith 1999, 75).

The overarching power of texts in this argument does not necessitate written texts. As other worlds have noted, though many of the texts in game development are undocumented, they are still enacted through daily "operating" practice. Daily practice can be taken as a text, as this work has attempted to construct. Developers who, as Smith notes, "enter and participate in them" at work also adhere to them at play. Yet the fact remains that, as with technology, there is a dialogue in which changes to software/code also back up to the production structures. At a more fundamental level, this textual mediation seems to be part and parcel of the normal process of technological development. Ruling relations, then, and the available avenues of action are dictated by the codes in place within technological production processes, and the codes themselves dictate what changes can be made. Software/code is the glue that provides corporate technology purveyors a way to make technologies that are themselves capable of changing over time. As users appropriate technologies, software/code allows for the very "text" of these technologies to be rewritten to rein users in.

Put in a different context, software/code becomes another assemblage by which the state will attempt to stop the lines of flight that "the abstract machine of mutation," will continually enable (Deleuze and Guattari 1987, 227). At the same time, those very systems are subject to mutation and it is the "impotence" of those that mobilize the state to trap this system rendering it a futile quest in which the state often unleashes the war machine. Software/code pushes the concept of ruling relations, because of the ability for these texts to shift over time, making user's agency in the network difficult to pin down, as it shifts according to changes in extra-local texts. Thus, software/code realizes a solution to one of the biggest "problems," in the network of technological production, those that prove difficult to hold in place:

> The network approach stresses this by noting that there is almost always some degree of divergence between what the elements of a network would do if left to their own devices and what they are obliged, encouraged, or forced to do when they are enrolled within the network. . . . Elements in the network prove difficult to tame or difficult to hold in place. Vigilance and surveillance have to be maintained, or else the elements will fall out of line and the network will start to crumble. (Law 1989, 114)

John Law indexes the particular desire, on the part of actor-networks, for divergence that causes me to question if software/code is not precisely the kind of glue that was previously fundamentally absent and that provided corporate technology purveyors the capacity to swerve as much as their users do. Because technologies can shift, change, and swerve, they are also capable of encoding the network approach within their processes and thereby become a hegemonic force, themselves. It is the possibility of shift, change, and swerve that makes technologies capable of being part of hegemonic discourse. Their ability to adjust to users and their appropriations over time is precisely what makes them difficult as compared with and different from previous generations of technologies. "Domination" in the technological arena becomes less about stationary stability and instead about the stability of particular texts over time and shifting fields of user activity. "At an accelerating pace in the twentieth century, the ruling relations come to form hyper-realities that can be operated and acted in rather than merely written and read" (Smith 1999, 84). Malleability, then, seems the key to the durability of extra-local texts.

This world points to and examines some of the rules and structures that undergird the videogame industry. It draws upon the instrumental gaming of my field research. Within game development production there are numerous technological and social apparatuses, which structure the rules that developers experience so acutely in Worlds 1–3. In the rest of the levels of World 7, I take on the role of the power gamer as a means to seek out structural conditions or systems that sit behind the worlds of game developers. I do so in the hope that more developers find the desire to understand what makes the game industry tick in the fashion it does. This first level has situated us in the technology world's need for domination, conflict, ruling relations and their texts, and the network approach to hegemony so that we can create an open space in which to interrogate the systems and structures documented in previous worlds.

The desire for and work of pursuing those underlying systems and structures is key to the investigative process. While there are surely technological mechanisms that legislate this game space, this world considers how technological components are networked with interested parties and political and legal structures of regulation to ensure enforcement in ways more effective than the simplest or most elaborate technological fixes. As we shall see, the force of these rules is truly felt when they come in concert.

World 7-2: Production Protection

As discussed in World 6, the intention of production control, at least initially, was to control the quality and supply of games entering the market, as well as to supplement the costs associated with selling console hardware systems at a loss. The lack of production standards and access to the production and distribution resources allows manufacturers to prevent "undesirable" material from being played on their consoles. Locking out independent developers provides the opportunity for brand management. It also engages the state's legal resources by using the ratings-system barrier to force companies into the industry's production control structure.[1]

In addition to controlling production, console manufacturers control distribution because they execute final quality assurance and testing of games and thereby exert direct control over all games playable on their consoles. This reality stems from a concern for maintaining the image of the console's brand rather than care for the kinds of games being positioned on the console. But the key here is that distribution and production have become entangled in the current game industry regime. The distribution networks have been disciplined by patent and legal structures. Thus, quality control disguises complete control. The game industry has always been a walled garden.

The tight control over production seems problematic because clearly low quality games make their way onto consoles and high quality games are blocked from entry. In response to this disconnect, Microsoft, one of the three console manufacturers, has opened up a partial production path to the public. A strong community has risen up in this nascent space, and developers have begun to share tools and practices more broadly than happens at established game development studios. Unfortunately, this breech in control of production still requires developers to focus exclusively on Microsoft's Xbox 360 console. In opening up game development on their console, they have closed off the possibility of opening up those paths across console devices, tapping into the pent up demand for more open access to game console development tools, while simultaneously ensuring that those tools can be brought to bear primarily only on their own consoles.

Developers gaining access to the console manufacturers' networks must frequently take a game that is already significantly developed and attempt to move it to consoles. A game developed for one platform is not necessarily easily ported to another platform, and it becomes impossible for a development team to account for, in advance, the idiosyncrasies of each platform. The specificities of production for those platforms remain closed, open only to the select few authorized by the manufacturer or publisher. Even when a developer is authorized to move a game to a new hardware platform, individual technological vagaries make a massive difference, as a Lost Toys Studio's engineer notes.

Our code structure was aimed toward making the porting process as painless as possible, but we hadn't counted on the extent of the limitations of the console platforms relative to the PC. . . . The Xbox port of the game had the advantage of being based on DirectX, and hence the majority of the code was shared with the PC version. The PlayStation 2 port, however, required an entire graphics and sound engine to be coded from scratch—a mammoth task for our two PlayStation 2 programmers, one of whom had never actually written any code for the machine before this project and was still supporting a significant amount of code on the PC tool-chain and Xbox sides of the project. (Carter 2003, 56)

Ultimately, the porting process impacts developers more than publishers or manufacturers. In the end, developers must re-create what they have already made, or face the retrospective knowledge that they could have accounted for limitations were they not coding under significant constraints made by corporate entities. Every game development studio must create processes that are then are re-made and re-learned constantly throughout the process of development. Developers frequently talk about processes that are not unique to their game, like "baking" data, a process that in many respects could be reasonably standardized among studios and across platforms, but has not.

The content baking process for the console was time-consuming and difficult to troubleshoot. Frequently the only way to either identify or resolve a bake problem was to re-bake at the cost of up to an hour of work, and if the tools were actually broken in some way, it would take at least another bake cycle to be able to work effectively again. (Finley 2007, 26)

As evidenced by the lack of documentation and standardization, and by the wastes of time and energy in every project that begins in the dark without the support they need, publishers and manufacturers make no effort to encourage sharing or collaboration across the industry, even across studios they own and manage. Manufacturers have no incentive to make the process of game development flow more smoothly. Too many developers hoping to breach the gates of access networks are throwing themselves at publishers and manufacturers. And for those established developers who've already gained access, the risks of mobilizing efforts on their own include criminal violation of the DMCA, violation of NDAs, and being cut out of the networks that allow them to work in these spaces. So demands for resources from without and within are blocked by those very corporations that stand to gain from better-equipped developers.

The entanglement of production and distribution significantly limits the game industry and ultimately affects the working lives of game developers worldwide. Sharing and collaboration take a back seat to simply keeping a development studio alive and functioning within the limits placed on production. On the other hand, when accusations of copyright infringement are true, frequently only force can dissuade individuals perpetrating theft, which is why companies are continually looking for new mechanisms to use the state to force what legal maneuvering seems incapable of doing: stopping copyright violations. In particular, corporations have come to desire the prerogative capacities of the state to ensure that their networks of access and secrecy are not compromised. Media corporations in particular have become quite adept at mobilizing state forces to meet their needs rather than the interests of users and producers, who are left to work within or break these assembled structures.

In its protection against unauthorized circumvention, the DMCA does much more than protect digital copyright; it will be the guardian at the gates of the trusted system, ready to patrol the boundaries of this massive control mechanism. And by emphasizing access rather than copying, it can sanction violations of the trusted system that have nothing to do with copying, but are rather about accessing materials without following the proper channels, i.e. paying for it, and following the rules prescribed by that commercial relationship. CSS and the trusted system proscribe behavior in intense detail and design other behaviors out of existence, and then depend on the law to ensure that consumers use the system as recommended, risking the threat of criminal penalty if they attempt otherwise. (Gillespie 2004, 224)

Domination, then, emerges not simply from the bottom up. Behavior is both deemed criminal or deviant by the game industry, as illustrated in figure 7.2, and regulated by the state. Simultaneously, software/code/silicon/hardware then encode those ruling relations in a way that enables them to be modified or updated to ensure that users, even when able to deviate, can be brought back in line with ruling relations via updated technological legislation. Domination is a product of both a top-down restrictive text of acceptable behaviors and a bottom up acquiescence to how things are assumed to "naturally" occur, which are influenced by those rules instituted from above, which emerged from bottom up practices that migrate ever upward. Domination is a mangled dance.

Some companies take this particular copyright interface between the state and corporation very seriously: Nintendo is one such example. Recently they have come out in strong support of the US stance on antipiracy measures. But there is an implicit conflation of actual illegal activity and potential illegal activity, and certainly no mention of the legal and legitimate uses of so-called piracy technologies, such as homebrew game development or listening to music on the devices. What makes these efforts troubling is that like all coercive state-based efforts, they suffer from the

GAMEDAILY BIZ

Top Stories
Classifieds
Industry Events
Daily Charts
Daily Polls
Features
News Articles
Ad Watch
My Turn
GD Interviews
Chart Toppers
Media Coverage
Industry Movers
Subscribe
Post Your Classified
Advertise Today
RSS Feeds XML
Contact Us
Send Us News Tips

Poll of the Day

Has your Xbox 360 scratched and ruined any of your game discs?

Find Articles Search

Industry News 6/26/2007 6:54:18 PM

ESA Applauds San Diego Raid

Notorious game pirate now faces 10 felony-level offenses

The Entertainment Software Association (ESA) congratulated the San Diego County District Attorney's Office and the Computer and Technology Crime High-Tech Response Unit (CATCH) for their recent raid and arrest of a game pirate in the San Diego area. Frederick Brown of Vista, California was arrested after CATCH officials conducted a search-and-seizure operation on June 13, 2007.

"We commend the San Diego County District Attorney's Office and CATCH for their efforts on this case as another example of their strong commitment to combating intellectual property theft," said Michael Gallagher, president of ESA. "The entertainment software industry will provide whatever support is needed for a successful prosecution and will continue to support law enforcement efforts to reduce game piracy in its many forms."

Brown allegedly had been running his pirate operation via online advertisements for years, and thousands of pirated discs and mod chips were seized in his home. He is currently facing 10 counts of felony-level offenses, including grand theft, computer crime and trafficking counterfeit marks.

by David Radd

Figure 7.2
A screen shot of San Diego Piracy Raid Report (Radd 2007)

inability to determine "the difference between false deference and real deference . . . how can we distinguish compliance under force from mystification and fatalism?" (Scott 1976, 230). The growing mobilization of the prerogative state power on the part of corporations is impressive, as the following game industry news report indicates, with new "processes" in which corporations offer their input on the "adequacy" of state intervention in the space of intellectual property rights enforcement : "Each year Nintendo participates in the annual Special 301 process, by which the US Trade Representative office solicits views from the industry and makes judgments about the adequacy of intellectual property laws and enforcement in foreign countries, including not only China, but Hong Kong, Brazil, Mexico and Paraguay as well" (Dobson 2007).

This power, ceded to game industry giants by the US government represents an important example of how the extra-legal aspects of the state are being mobilized by corporations to police individuals. This can be seen in recent controversies around the Stop Online Piracy Act (SOPA) and Protect

IP Act (PIPA). Our technologies are allegedly advanced enough that legislators and law enforcement find themselves at a loss for what should be done, and thus cede power to industry. While companies argue and lobby for decreased regulation of themselves, indicating the importance and intelligence of the market's ability to solve problems, they simultaneously lobby for increased regulation of the individual user/citizen. In many cases these are military-like raids on homes and businesses, which fail to even meet the goals of the companies that mobilize these interventions.

> Nintendo commented in a statement: Despite the millions of counterfeit Nintendo products seized from retailers and manufacturing plants in China through the years, there has only been one criminal prosecution. Numerous factories, where tens of thousands of counterfeit Nintendo products were seized, escaped with only trivial fines or no penalty at all. And often these production sites continue to operate after products are seized. In order to avoid punishment, many counterfeiters are sophisticated and keep stock levels below the criminal thresholds and avoid keeping sales records. (Dobson 2007)

These contradictions weaken the game industry's claims that market-based ratings systems like the ESRB are effective. In the case of content regulation, allow the market to solve the problem. In the case of copyright and piracy, legislate the solution. If the game industry cannot determine whether the market or the state should be the dominant mode of disciplining their production methods, they subvert their own claim that only a market-based solution to the consumer's understanding of games ratings. While criminalizing the sale of mature- or adult-only-rated games to minors seems a foolhardy approach to managing the situation, the game industry and the ESRB have not been particularly effective at coming up with alternative approaches to how to address the problem. The criteria and varying moralities at play in the ratings of games ultimately goes unquestioned: "I take issue with the fact that to get an E10 rating, we had to change our beloved secret bonus character's name from Armondo Gnuetbahg to Armondo Ootbagh. Am I missing something here, or is Gnuetbahg a new curse word that the hip kids are using? Yes, in our tutorial we teach kids to "clobber 10 people before time expires" but we aren't allowed to say 'Gnuetbahg'" (Schadt 2007, 34).

As time has gone by, the situation has become more complex. With the introduction of the DMCA and the use of encryption schemes in console game systems, attempting to circumvent the limits on production has become a criminal activity (DMCA 1998). While the DMCA has come under particular scrutiny recently because of its relationship with digital rights management (DRM) in the context of digital movie players, there has been very little broader public scrutiny of the practice and legislation. Legislators, unable to understand the complexity and nuance of the industry and work that occurs within it, abdicate their responsibility and pass legislation embraced by individual companies, but often not by those responsible for creating games. In many respects DRM technologies are actually a collective invention of the videogame industry. Users interested in playing music files on different devices seem to have gotten the bulk of the attention[2], yet game content has long been restricted and many users fail to see their rights extend to game content. While this is good for those already in positions of power in the videogame industry, one might posit that the inconsistency or inattention to this detail is especially problematic because it erodes the foundations of media production companies' arguments. Some in the videogame industry have noted this contradiction, though they seem unsure what the consequences of that variance means precisely (Fahey 2007). The "postmodern" state is not one in which the state is strictly on the decline—particular components are on the decline and others are in ascendance, and particularly those dealing with restrictions: the police, surveillance, and military action.

> The [s]tate, constituted as a coercive system of authority that has a monopoly over institutionalized violence, forms a second organizing principle through which a ruling class can seek to impose its will not only upon its opponents but upon the anarchical flux, change, and uncertainty to which capitalist modernity is always prone. The tools vary from regulation of money and legal guarantees of fair market contracts, through fiscal interventions, credit creation, and tax redistributions, to provision of social and physical infrastructures, direct control over capital and labour allocations as well as over wages and prices, the nationalization of key sectors, restrictions on working-class power, police surveillance, and military repression and the like. (Harvey 1990, 108)

This claim that the secondary role of the state answers to a ruling class evokes Gillespie's assertion that the coercive arm of regulation is both creating and enforcing boundaries for the game industry.

Recently the same mechanisms have been deployed in an unprecedented manner. In tandem with the US Immigration and Customs Enforcement agency, the game industry has leveraged the punitive powers of the state to execute search warrants in 16 states across the United States in a preemptive strike against equipment that *could* be used for piracy. Despite the possible legitimate uses of technology, the risk of possible illegal activity becomes the motivation for these raids. The scale of these recent actions are of particular note, verging on militarized rather than localized police actions.

Illicit devices like the ones targeted today are created with one purpose in mind, subverting copyright protections," said Julie L. Myers, Assistant Secretary of Homeland Security for Immigration and Customs Enforcement. "These crimes cost legitimate businesses billions of dollars annually and facilitate multiple other layers of criminality, such as smuggling, software piracy and money laundering. . . .

Between fiscal years 2002 and 2006, ICE agents arrested more than 700 individuals for IPR violations and dismantled several large scale criminal organizations that distributed counterfeit merchandise to nations around the globe. At the same time, ICE investigations into these networks resulted in 449 criminal indictments and 425 convictions. Together, ICE and CBP seized more than $750 million worth of counterfeit goods from fiscal year 1998 through fiscal year 2006. (ICE 2007)

The lack of response by active game developers, or potential developers more broadly indicates that these activities are seen as either normal or not even worth paying attention to. Yet, what is being seized in these raids is a device of potential misuse that also serves very real and legitimate uses as well. Part of the market appeal of these devices is that they enable illegal activity on the part of users. They are also appealing because they expand the legitimate uses of the devices as well. What is being described as a device solely costing companies money is not necessarily costing them anything. In many respects, these devices actually expand the capability of a users' technological investment in ways that are not approved…just because they're not approved. These devices make possible the kind of DIY and remix cultures that are so critical to new media industries. Yet, it is normal for the state to break into houses to take these devices, and thus no counter argument is made. The lack of debate itself, though, might recapitulate resistance: "What passes as deference 'is ritualized and habitual' or even calculating. . . . There may in fact be a large disparity between this constrained behavior and the behavior that would occur if constraints were lifted. The degree of this disparity would be some index of the disingenuousness of deferential acts. The very act of deferring may embody a certain mockery" (Scott 1976, 232).

In many respects this mocking nod to adherence of form is evidence among a new generation of media producers, those with an eye toward ironic re-interpretation or "remixing" (Lessig 2005). This new sense of playful disregard for (even while technically adhering to) constraints mimics that of other technical art forms and may follow their trajectory. As more users understand themselves as capable producers, the current legitimacy of this state regulatory action, particularly preemptory silencing of potential legal creation will be called into question, as has already been demonstrated in areas like digital music downloads (Gillespie 2006).

The tension between the fixity (and hence stability) that [S]tate regulation imposes, and the fluid motion of capital flow, remains a crucial problem for the social and political organization of capitalism. This difficulty is modified by the way in which the [S]tate stands itself to be disciplined by internal forces (upon which it relies for its power) and external conditions—competition in the world economy, exchange rates, and capital movements, migration, or, on occasion, direct political interventions on the part of superior powers. (Harvey 1990, 109)

The DMCA has further extended the ability of corporations to incarcerate people who, regardless of their intentions, enable others to circumvent those copy protection mechanisms that companies create. The legal (though inaccurate) conflation of "hacking" and "cracking" derives from the tension between the rights of users to legally do with technology as they please ("hacking") and restriction against users illegally attempting to copy or redistribute the property of corporations ("cracking"). This conflation renders the DMCA problematic because its foundations are rooted in the assumption that anyone interested in doing something with digital data that it was not originally intended to do is attempting to make illegal copies. The state is making no distinction between hacking and cracking. Yet the same limitations extend to anyone interested in producing media for console videogames. There is no acknowledgment that these technologies limit producers within structures that should not apply to them. Production of new media is only talked about vaguely as "home brew" and is both denigrated and extolled in different industry press releases. The erroneous assumption of illegality is encouraged by companies precisely because it gives them power to enforce their existing technologies and legal structures with new regulations and police actions not in the interest of users.

World 7-3: Patent and Copyright "Risk"

Patents and patent enforcement, which severely hems in gaming competition, are the legal structures that support major corporate players in the game industry. One might argue that growth of the industry faces a significant problem when corporate SEC filings begin to indicate the litigious character of existing US copyright and patent systems as a key risk factor.

If patent claims continue to be asserted against us, we may be unable to sustain our current business models or profits, or we may be precluded from pursuing new business opportunities in the future.

Many patents have been issued that may apply to widely-used game technologies, or to potential new modes of delivering, playing or monetizing game software products. For example, infringement claims under many issued patents are now being asserted against interactive software or online game sites. Several such claims have been asserted against us. We incur substantial expenses in evaluating and defending against such claims, regardless of the merits of the claims. In the event that there is a determination that we have infringed a third-party patent, we could incur significant monetary liability and be prevented from using the rights in the future, which could negatively impact our operating results. We may also discover that future opportunities to provide new and innovative modes of game play and game delivery to consumers may be precluded by existing patents that we are unable to license on reasonable terms. . . . Other intellectual property claims may increase our product costs or require us to cease selling affected products.

Many of our products include extremely realistic graphical images, and we expect that as technology continues to advance, images will become even more realistic. Some of the images and other content are based on real-world examples that may inadvertently infringe upon the intellectual property rights of others. Although we believe that we make reasonable efforts to ensure that our products do not violate the intellectual property rights of others, it is possible that third parties still may claim infringement. From time to time, we receive communications from third parties regarding such claims. Existing or future infringement claims against us, whether valid or not, may be time consuming and expensive to defend. Such claims or litigations could require us to stop selling the affected products, redesign those products to avoid infringement, or obtain a license, all of which would be costly and harm our business. (Electronic Arts 2007, 54)

"Risk and uncertainty" caused by the current condition of the patent and copyright institutions in the United States points to a fundamental problem for the future of videogame production and for new media production more broadly. In many respects the risk and uncertainty is exacerbated for those companies who, unlike Electronic Arts, do not have the available capital to defend themselves from the numerous copyright and patent claims that could conceivably be brought against them. Despite the problematic character of intellectual property law in the United States, it remains the standard on a global scale. Even more troubling, new legislative efforts funded by corporations push hard against ideas like fair use or a creative commons, which will only create more risk and uncertainty for new media producers.

Copyright and patent infringement claims against established companies in the videogame industry are only an index of a more substantial problem. These companies have the money and experience that allows them to be more able to deal with infringement claims. Smaller and newly created development studios looking to establish themselves actually have more to lose in this environment than companies like Electronic Arts, who shine a light on the problem. Ultimately copyright was designed to "promote the progress of science and useful arts" (Sprigman 2002), of which it has done very little for the videogame industry broadly.

The volley of infringement claims is so pervasive that several game industry lawyers have indicated in conversations with me that new studios need to assume that within their first year of activity, they will be taken to court by one corporation or another. Beyond being part of the pervasive reality of legal and corporate teams, at a practical level, copyright and patent infringement claims have come to impact the daily lives of game developers. A pervasive environment of conservatism surrounds the legal analysis of videogames. The assumption is that if a patent or copyright might apply, then it ought to be preemptively licensed, purchased, or the game altered. The lead designer of the game *Tony Hawk's Downhill Jam* for the Nintendo Wii talked about dealing with corporate legal teams during development work.

> We also had a number of changes to make due to a fear of potential lawsuits. This exchange is a prime example:
>
> Activision Legal: "You'll have to change that restaurant's name."
> Development Team: "But it's called Dim Sum. That's, like, totally generic."
> Activision Legal: "But if you type Dim Sum into Google, you'll find many actual, real-life restaurants called 'Dim Sum.' It's safer just to change it."
>
> I failed to realize how frightening the legal climate is at present. The fear of being sued is so pervasive that artistic freedom is being compromised, and conservative, safe decisions are routinely made even when there is no legitimate legal infringement. (Schadt 2007)

Companies' use of copyright and patent to strictly police their intellectual property encourages this conservatism, which then ensures the continued failure of copyright and patent law to promote "progress" and instead encourages regress. As more and more allowances are made, claims of fair-use and public domain knowledge are diminished, based simply on fear of litigation. Numerous patent and copyright infringement claims are filed with the knowledge that companies will be more likely to pay off the claimant than actually attempt to fight or correct the broader problem. This is in part because those same defendant companies have a vested interest in being able to pursue similar litigation against other studios.

Console manufacturers also make significant use of the slippery slope of copyright and patent law to control production and distribution systems. Copyright claims are a mechanism for shutting down retailers of "pirate" technologies. Those same technologies, however, are integral to truly independent game development: game creation entirely outside of the videogame industry's networks. Again, the ability to control distribution (copyright) becomes conflated with the ability to produce (speak). As such, many of these attacks on "copyright infringement" need to be reframed as attacks on speech by those who are being silenced. The ability for people to learn, investigate, and share information about these devices via mechanisms that game companies do not control is not purely attributable to piracy or violations of copyright. In many cases attempts to share information are simply attempts by the user/producer to work on these new devices. In light of this, game developers are those most hurt by litigious abuses, and should seek reform. They have a particularly interesting understanding of the complex connections among the creative interdisciplinary work necessary for the production of new media, and should involve themselves in the process of copyright and patent reform, reminding lawmakers that legal decisions ultimately impact workers/voters. Game developers need to help a broadly defined public understand the importance of these issues, especially in connection with their ability to think with and comment on cultural forms. In a world now characterized by technologies that blur the lines between user and producer, the ability to investigate, experiment, and tinker is especially important. The realities of "Web 2.0" or "Participatory Culture" are endangered by the continued attacks on the very core mechanisms that support it. For game companies already at the center of the industry, software/code and supporting structures that mobilize the state support the stability of those positions. Yet, this mobilization is anathema to new emerging forms of media and precisely those communities of practice from which game development emerged. To say that the structure of the game industry is a bottom-up process neglects the ways in which powerful and power-laden structures are being put to use to further structure the way the game industry functions. These new structures favor the status quo and discourage innovation and information sharing among game developers. Quite literally, the game industry has structured itself and continues to fight to maintain a structure that hurts game developers.

World 7-4: The Death of Hacking and Homebrew

As mentioned in the first level of this world, homebrew game development is independent game development for console game systems, like the Nintendo Dual Screen (or "DS") or Sony's PlayStation Portable (or "PSP"). Traditionally, these devices are covered under significant numbers of user-end license agreements aimed at users. They demand that the devices be used in only particular ways. To produce software for these devices, one must be a registered or "licensed" developer for Sony or Nintendo. This license, combined with not an insignificant sum of money, provides developers with the requisite hardware and software necessary for developing software for these systems.

Homebrew game development circumvents the restrictive elements that prevent users from creating software for these systems. As an informant writes: "The point of homebrew is to allow programmers who want to develop something for the Wii [or the DS, or the PSP] to have the chance . . . Frankly, as much as I love Nintendo, the fact that they make it impossible for people to program for the Wii without having a company of X size, Y wealth, and Z experience created the need for the homebrew channel" (Informant 2009a).

This person's argument is intriguingly rooted in a kind of market logic. Nintendo's lack of attention to the numerous hopeful game developers and tinkerers created a demand for circumvention devices. These emergent counter-channels are a product of the restrictive policies. For the majority of would-be developers, if a legitimate means were available, even for a small fee, that route would be explored first, rather than the more complex illegitimate means.

As an experiment, a game, a homebrew test of my own, I have imported from China numerous "copyright circumvention devices," like those seized in ICE's raids across the United States. I have imported devices created for the Nintendo DS. I import more of the "MOD Chips" than I need and more than I'll ever use. I do it now almost out of spite, as new devices are developed I purchase several at a time. It has become a kind of game for me, wondering when I'll pull the slot-machine lever and get a surprise that I hadn't intended: federal agents at my door. Part of me wonders if this confession will result in my own disciplining. If this isn't a "chilling effect," then I am not sure what is. I have never used these devices to illegally violate copyright and play games that I have not legitimately paid for. I have used these devices to allow me to develop small games for the DS and make presentations using the device at academic conferences. This legitimate activity, which fundamentally should be protected by the first amendment and fair use, has been criminalized at the demand of videogame manufacturing companies.

It is the "extra legal" aspect of the state—its ability to incarcerate and ultimately punish or kill individuals—now being leveraged by corporations hoping to cement their commercial interests. Press releases notify the general public of the state's activity and in most cases there is little or no public outcry, media coverage, or discussion of what might be mistaken in these situations. There are indeed legitimate and illegitimate uses of the devices being taken so seriously by media producing companies. This is little different than the range of legitimate and illegitimate uses for products like photocopiers, CD/DVD burners, computers, or simply pens and pencils.

In response to this unjust abuse of state power in the interest of corporate profit, entire communities of software developers have formed (composed largely of engineers), to support the development of software fundamental to the game development process. At the same time however, these software systems are dependent on numerous technologies, processes, and hardware/software glitches that make it possible for them to load their software onto devices that are otherwise locked.

A side effect of this is that the act of making these devices open to would-be developers means that they create new openings for software "piracy." Confounding the efforts of developers who want to create are those who are interested in cheating the manufacturers. Once devices are unlocked, ROM files are traded among those with little interest in making software for the devices but with significant interest in playing all the games available for these systems without payment. Others focus on creating game-emulation software, allowing their PSP or DS to play games that were never intended for it. This, probably more than anything is what raises the ire of console manufacturers. They would prefer that you not dilute their brand by running *Super Mario Bros.* on your PSP or that if you were planning to play *Sonic the Hedgehog* on your DS, they would prefer you pay for it again, despite having purchased it for the Sega Genesis back in 1991. So rather than being a purely free speech, game development issue, there are the realities of piracy, brand dilution, and others: "Doesn't the whole idea behind homebrew revolve around piracy? I see a bunch of topics involving emulation . . . Isn't that piracy? . . . I don't understand the whole Puritan approach. I guarantee that anyone who has homebrew on their Wii has at least one pirated IP" (Informant 2009b).

And while I cannot disagree more—to my mind emulation is not piracy—it often does quickly become that. Most people interested in emulation do not own licenses to the software they wish to emulate. But that does not mean that there aren't significant numbers of users with legitimate claims to emulation.

Because of the piracy and abuse of technology, these homebrew platforms quickly become the target of the powers that be. Nintendo and Sony release software updates to their consoles disabling homebrew efforts.

Sony has even gone so far as to remove capabilities from the PS3 that once allowed it to run Linux, an Open Source Operating System. The PSP frequently undergoes firmware updates that disable previous generations of homebrew efforts, only to days later be re-"hacked," restoring homebrew developers' freedoms.

The DS is a strange bird in this mix of firmware updatable console systems, a mistake Nintendo has "corrected" with their release of the DSi, a camera enabled, SD-card-packing downloadable game system. This system contains flash-able firmware. But the DS does not have an updatable firmware. However, this does not mean that Nintendo lacks the urge to prevent the production of unauthorized software.

To correct for the DS's lack of updatable firmware, Nintendo has mobilized the extralegal aspect of the state. Multi-state, multi-site, simultaneous raids of US businesses selling R4 or DSTT cartridges that bypass the DS's encryption system have become more common. Nintendo has made appeals to the US government, the Chinese government, and others to stem the flow of these devices (ICE 2007). This, more than anything troubles me as a social analyst, the mobilization of the extra-legal aspects of the state for interests that are largely market concerns. While the game industry will simultaneously defer to the power of the market to regulate content concerns or the ability for the state to censor game content, they quickly request the state's ability to incarcerate and seize property.

To counteract these limitations, users have been willing to open up their consoles and install "Mod-Chips," which have three effects. One allows a user to legally import a game and play it (circumventing a regional lockout), and a second that allows users to illegally "burn" or copy game disks. A third and less publicized effect is access for those interested in developing games for consoles outside the game industry's networks. Truly independent game development, work disconnected from the networks of secrecy and access requires breaking current US intellectual property rights laws. In the early part of the century, companies who offered services to assist users in the modification process were subsequently taken to court by console manufacturers for violation of copyright (Nintendo and Lik Sang 2003; Sony Computer Entertainment and Lik Sang 2003). Strikingly, in these cases the copyright of the console manufacturers was not being violated, but their mechanisms for control *were* being circumvented. The issue is even foggier in countries like India where a particular console may not be available in the first place. What region is the user a part of? Console manufacturers assert that if a user imports a game console to play, they must ensure that all games they buy subsequently are from the proper region.

In many respects I see MOD or remix culture (Lessig 2005) embodied by these activities of hackers to be a serious complication to any notion of technological systems as "durable" (Latour 1991) in any kind of lasting way without connections to systems that enforce durability seems problematic. Of course barriers are in place that encourage durability or stability, yet an increased interest by users of technology in MODing or remixing them increases the complexity of the situation.

In a well-known case among game developers the site "Lik-Sang" was shut down because of continued harassment from console manufacturers, despite their service to the game development community. Lik-Sang was the leading provider of adapters for console controllers so that developers could use them on PCs during the process of development. Some of the products were even being used by smaller development studios to supplement the number of available DevKits within their company. This case is also indicative of a broader problem of litigation practice, where companies with more money, despite possibly having an invalid case, can inflict mortal financial wounds on those companies that seek to interrupt or alter networks of access.

Lik-Sang.com Out of Business due to Multiple Sony Lawsuits
Tue Oct 24 2006 21:58:51 Hong Kong Time—Corporate Info
OUT OF BUSINESS NOTICE

Hong Kong, October 24th of 2006—Lik-Sang.com, the popular gaming retailer from Hong Kong, has today announced that it is forced to close down due to multiple legal actions brought against it by Sony Computer Entertainment Europe Limited and Sony Computer Entertainment Inc. Sony claimed that Lik-Sang infringed its trade marks, copyright and registered design rights by selling Sony PSP consoles from Asia to European customers, and have recently obtained a judgment in the High Court of London (England) rendering Lik-Sang's sales of PSP consoles unlawful.

As of today, Lik-Sang.com will not be in the position to accept any new orders and will cancel and refund all existing orders that have already been placed. Furthermore, Lik-Sang is working closely with banks and PayPal to refund any store credits held by the company, and the customer support department is taking care of any open transactions such as pending RMAs or repairs and shipping related matters. The staff of Lik-Sang will make sure that nobody will get hurt in the crossfire of this ordeal.

A Sony spokesperson declined to comment directly on the lawsuit against Lik-Sang, but recently went on to tell Gamesindustry.biz that "ultimately, we're trying to protect consumers from being sold hardware that does not conform to strict EU or UK consumer safety standards, due to voltage supply differences et cetera; is not - in PS3's case - backwards compatible with either PS1 or PS2 software; will not play European Blu-Ray movies or DVDs; and will not be covered by warranty."

Lik-Sang strongly disagrees with Sony's opinion that their customers need this kind of protection and pointed out that PSP consoles shipped from Lik-Sang contained genuine Sony 100V-240V AC Adapters that carry CE and other safety marks and are compatible worldwide. All PSP consoles were in conformity with all EU and UK consumer safety regulations.

Furthermore, Sony have failed to disclose to the London High Court that not only the world wide gaming community in more than 100 countries relied on Lik-Sang for their gaming needs, but also Sony Europe's very own top directors repeatedly got their Sony PSP hard or software imports in nicely packed Lik-Sang parcels with free Lik-Sang Mugs or Lik-Sang Badge Holders, starting just two days after Japan's official release, as early as 14th of December 2004 (more than nine months earlier than the legal action). The list of PSP related Sony Europe orders reads like the who's who of the videogames industry, and includes Ray Maguire (Managing Director, Sony Computer Entertainment Europe Ltd), Alan Duncan (UK Marketing Director, Sony Computer Entertainment Europe Ltd.), Chris Sorrell (Creative Director, Sony Computer Entertainment Europe Ltd.), Rob Parkin (Development Director, Sony Computer Entertainment Europe Ltd.), just to name a few.

"Today is Sony Europe victory about PSP, tomorrow is Sony Europe's ongoing pressure about PlayStation 3. With this precedent set, next week could already be the stage for complaints from Sony America about the same thing, or from other console manufacturers about other consoles to other regions, or even from any publisher about any specific software title to any country they don't see fit. It's the beginning of the end . . . of the world as we know it," stated Pascal Clarysse, formerly known as the Marketing Manager of Lik-Sang.com.

"Blame it on Sony. That's the latest dark spot in their shameful track record as gaming industry leader. The Empire finally 'won', few dominating retailers from the UK probably will rejoice the news, but everybody else in the gaming world lost something today." (Lik-Sang.com 2006)

As pervasive as copyright litigation is, patent litigation is much less common. Patents in the game industry are primarily used to prevent other companies from using the same methods, techniques, or technologies as the patenting organization. When any mechanism of control is circumvented, copyright violation is often the first rallying cry, even when it isn't clear that actual violations have occurred. Thus, it is often the catch-all weapon in shutting down unauthorized production. Even when the accusation of copyright infringement is false, often the individual or company being sued will attempt to placate the attacking corporation. Taken to its eventual conclusion, this technique of instantaneous and often unjustified copyright policing often leads to endings like that which befell Lik-Sang.

Beyond the implications of putting companies out of business with spurious lawsuits, game corporations need to be challenged for locking out independent voices. Aspiring, independent, hobbyist, and student game developers also need to assert their rights to speak on devices that have been shut off. The ability for game developers to speak is being shut down through the over-application of patents and copyright. Co-opting the legal mechanisms of the state, console manufacturers have deliberately thwarted homebrew efforts by closing console systems to these developers. Game developers must assert that this is a violation of their speech and we must all resist the ways in which copyright and patent law have become a threat rather than a cultivator of new speech and speech forms.

Ultimately, changing the game industry's control over speech must occur at the policy level, and will require the activism of numerous game developers. There will be resistance from console manufacturers and publishers, and it is likely that these corporations will use the threat of network access as a means to prevent changes they mistakenly see as disadvantageous. This is when it becomes imperative that studio heads use their positions to push for change. A drive of this sort cannot succeed without broad industry support and collaboration. And hackers are chief among those who must correct the misperception that their involvement amounts to piracy. Most industry leaders continue to conflate the difference between hacking and cracking, using a single word to reference two very different activities, despite the benefits hackers have to offer the videogame industry. A game industry executive epitomizes the refusal to acknowledge the nuance between those activities that are actually crucial to their survival:

> "Unfortunately, hackers will try to exploit any hardware system software," SCEA spokesperson Dave Karraker told GamesIndustry.biz.
>
> "The best we can do as a company is to make our security that much stronger and aggressively pursue legal action against anyone caught trying to use an exploit in an illegal manner.". . . Every hardware launch brings with it a race for hackers to defeat the system's protections, whether for the technological challenge, to run copied software, or to allow for homebrew games.
>
> Despite Sony's attempts to prevent its emergence, the PSP has a strong homebrew community—and hackers are doubtless hoping to establish a similar base for PS3. (Androvich 2007)

What is the difference between exploiting a hardware system to do what you would like it to do, and trying to use it in an illegal manner? I would answer that this is precisely the distinction between the hacker and the cracker, yet companies continue to combat them as if they were the same entity[3], and apparently, producing for a console in an unlicensed manner fits this description. There is a huge difference, despite Karraker's refusal to see it, between homebrew software and illegally copied software. "Homebrew" software is developed by amateurs at home: aspiring students and developers hoping to learn about game development practice, as well as hobbyists hoping to tinker with the devices they have already paid large sums of money for. Though the same exploits may contribute to both hacking and cracking, it seems premature to pursue both as if they were the same thing. Some companies, like Nintendo largely ignore homebrew developers. Sony actively combats them. And Microsoft has ostensibly embraced them.

One story in particular stands out as an example of how homebrew development benefits the videogame industry. Nintendo's GameBoy Advance (GBA) handheld system has enjoyed one of the longest lives of any console game system, in part because of its relatively low cost and large library of videogames. More importantly, it developed a large community of homebrew developers who invested significant time and energy into making the system accessible and open to new developers. Even the Nintendo Dual Screen (DS), the logical successor to the GBA will play cartridges made for that system. GBA development in its later years benefited greatly from the development of an open source and homebrew project called VisualBoyAdvance,[4] a project that began as an emulator for the GBA on PCs. As the project matured, so did the development tools that the software included. Built-in map viewers, sprite viewers, memory viewers, palette viewers, and visual debugging tools were all integrated into one package by hackers seeking to make games for the device. Not even Nintendo had provided such a host of tools for developing games for the GBA—even the software libraries associated with GBA homebrew began to surpass those supplied by Nintendo. Licensed developers began developing tools with hacker-made software. While it is possible for someone to download the emulator and then download ROM files created from GBA cartridges, this is not the only possible use of the technology, and is a model for how operating outside the access network is productive and beneficial rather than being profit-usurping piracy.

Emulators, decompilers, and numerous other technologies that might be labeled as dangerous to the videogame industry are not obviously so. These technologies allow artwork generation and visualization creation. Each of these activities is based on technology that at one time was considered illegal by many in the videogame industry. If developers can make manufacturers and publishers see that this hacking is not the equivalent of cracking, all can benefit and independent voices will be heard.

The other complication to the hacker/cracker distinction is that as far as most companies are concerned, even if you do own the ROMs that you're playing on an emulator, they would rather be able to re-sell you that content than have you make use of it yourself. Having Nintendo's *Mario* on a Sony PS3 or Microsoft Xbox360 does not help videogame brand building initiatives or profits. Even "open" consoles, which have attempted to break down these barriers for developers, has economically failed because hobbyists put more development work into emulation than new games. In practice, emulation actually gives console manufacturers more reason to combat homebrew rather than embrace it.

World 7 Boss Fight: Is That Your Head Over There?

In the movie *The Princess Bride*, a character named "Miracle Max," played by Billy Crystal, entreats a client to "Have fun stormin' the castle." His wife, a moment later asks, "Do you think it will work?" to which he replies, "It will take a miracle." I am certainly no Miracle Max, and this is the videogame industry, not a book/movie. Yet the pessimism of the statement holds true, for there is more than one castle to storm and there are many levels.

Part of the difficulty of the current system stems from the compelling nature of the functionalist argument, "Yes, but if it is so broken, then why does it seem to work so well?" Despite all of the contradictions, the videogame industry broadly speaking has a compelling irrational stability and adaptability. Perhaps more importantly for some, it continues to bring in massive amounts of money. Why on earth would I want to storm such a formidable castle? The answer is of course part personal and part analytic. At an analytic level, I believe that the changing relationship between users and producers as examined in this world signify a significant schism for media producers. The rising use of coercive state power ought to be an indicator of this critical moment. I use the word coercion deliberately; drawing on the idea that hegemony is an ongoing process of coercion on the part of the state and consent on the part of those it governs regardless of if the consent is explicit or implicit (Gramsci 1975). If the game industry does not adapt to this changing relationship, it will not continue to exist as it does now, and much pain and strife will come to workers in the game industry before it comes to those in positions of power. There are also limits to the desire machine, which has driven the game industry for so long. As users become more capable of producing and fulfilling their desire to produce games of their own, the industry will not be able to sustain its churn-and-burn attitude toward employees. At a personal level, the answer is more esoteric. Just because something functions does not mean that it isn't capable of functioning in ways that are more respectful, nimble, and nurturing. I can only hope that a commitment from those who believe videogames are an art form will encourage game developers to seek changes to their communities of practice.

So while the following SEC filing excerpt indicates the risks associated with working within these licensing agreements, it neglects to indicate the effect that these agreements have on the practice of game development more broadly. Perhaps more importantly, it fails to acknowledge that efforts like those around homebrew and jail-breaking could significantly reduce these kinds of risks. Yet, time and again, publishing companies assuage their interest in alternatives for access to the next shiny toy. As even Securities and Exchange Commission filings note, the position of development studios and even publishing companies is dramatically affected by the control of hardware platforms.

> The videogame hardware manufacturers set the royalty rates and other fees that we must pay to publish games for their platforms, and therefore have significant influence on our costs. If one or more of these manufacturers adopt a different fee structure for future game consoles, our profitability will be materially impacted.
>
> In order to publish products for a videogame system such as the Xbox 360, PlayStation 3 or Wii, we must take a license from the manufacturer, which gives it the opportunity to set the fee structure that we must pay in order to publish games for that platform. Similarly, certain manufacturers have retained the flexibility to change their fee structures, or adopt different fee structures for online gameplay and other new features for their consoles. The control that hardware manufacturers have over the fee structures for their platforms and online access makes it difficult for us to predict our costs, profitability and impact on margins. Because publishing products for videogame systems is the largest portion of our business, any increase in fee structures would significantly harm our ability to generate revenues and/or profits. (Electronic Arts 2007, 53)

Concern about cost structures over production practices is indicative of the fundamental disconnect the game industry needs to make between production practice and the secrecy that surrounds those devices. Instead of attempting to make significant modifications to these systems of relation, developers who have gained access to these production networks willingly trade their ability to share and learn for an opportunity to make more money and produce another game title. The continued dominance of the console game market, which represents the largest portion of the game development business, also symbolizes a critical point of access that must be examined. Even online distribution networks, which remove a significant aspect of the risk associated with publishing game titles, remain closed and unavailable except to those companies that work within the licensing agreements of console manufacturers.

Box 7.2

#: SET DEMO_MODE 1
AUTHOR_DEV_DILEMMA: For many of my informants, the independent game development "scene" has emerged as a kind of alternative to the games that dominate the mainstream game industry. In recent years, the independent game development community has come into its own. As shared tools, resources and open conversation about game design and development has enabled a flourishing, though somewhat insular, creative zone of game development.
Casey: Are there things going forward that are exciting to you?
DESIGN_LEAD_1: Yeah, why didn't independent game development come along earlier? There is more of a scene now, because the tools and the resources for making small innovative stuff has come down. There are avenues for distributing independent content to a mass market, which there wasn't before. I'm also interested in socially responsible games and serious games. Games still have such a limited audience. People tout all the time that it is bigger than the movie industry, blah, blah, blah, but they're only comparing box office sales with game sales. That doesn't include DVDs, merchandising, and everything else. Games need more avenues, not just the single one we have now.

Which is another thing. Gameplay length. It's one of the reasons games cost more, because it's supposed to be a longer experience. You have to create more crap. They charge what they do also because the precedent was already set. I guess it all comes down to, "Why can't we be more creative?" I guess my answer to that is, "Fine. I'll be creative when I don't have to worry about getting paid any more."
AUTHOR_DEV_DILEMMA: For so many of my informants, game development will only become sustainable when game development can extend into more areas, providing new opportunities for creativity that no longer demands adherence to the current industry structure. In the case of "indy" game development, that may mean sacrificing a salary, though alternatives also exist. Serious games, social games, and so-called casual games are each opportunities to create a more sustainable space of game development that is not so singularly focused on the current structure of the game industry.
#: SET DEMO_MODE 0

More broadly, the game industry must decide how it would like to position its relationship between the state and the market. The contradictory appeals to the independence of the game industry and the importance of the market is subverted by continued appeals to the state to provide protection and enforcement on activities that can arguably be defined as "piracy." The mobilization of the state as a means to enforce artificial controls on the market, which ultimately impact the everyday lives of game developers, is problematic. Those same legal activities severely constrain the ability of game developers to share and collaborate. This results in game developers spending significant amounts time reinventing the same components and never having the time necessary to share specific details about game development. While the double-speak creates particular opportunities for console manufacturers to manage and control the game industry, the same activities discourage any kind of maturation or the encouraging broader participation in game development practice.

In many respects the failure of the market and the state is visible in the kinds of games that are produced and distributed in the game industry. The continued consolidation and resulting risk-averse conglomerate companies encourages an adherence to the status quo. Only a small number of individuals within those corporations are given the freedom to push the game industry in new directions. Aspiring game developers are not given an opportunity to fully participate in the market of the game industry because they are constrained by the activities of established companies.

There seems to be something particularly distinct occurring in this space, something that I believe cuts to the core about why coercive state action is being mobilized on an unprecedented scale against entirely nonviolent citizen action. The (re)productive capacity of users rather than consumers is integral in this crisis for those seeking to control productive capacities. Play and playfulness is not something that corporations or the state has learned to work within yet, and as a consequence you find strong rhetoric against those that seek to leverage technologies to meet their desires rather than those they have been expected to consume.

Ultimately the relationship the industry has with the market and the state becomes a game of prisoner's dilemma. Will the players cooperate or defect? The unfortunate answer is that, as the game has currently been constructed, the players will tend to defect, only to have greater restrictions placed on them. The industry loses, because they, too, time and again defect. Yet, greater reward could be had if they had instead worked with users and learned to play something other than their own game.

Epilogue: The Videogame Industry Game

World 8: A Game Design Document

Box 8.1

```
#: SET DEMO_MODE 1
AUTHOR_DEV_DILEMMA: Never demo the end of a game.
#: SET DEMO_MODE 0
```

World 8-1: What's in a Game?

There is nothing like doing fieldwork among game developers to impress upon you that game design matters, and game mechanics are just as important as any other aspect of game design and development. I may have already been a convert. However, I grew up among the Nintendo generation, which is apparently enough to change one's perspective on the potential and possibilities for videogames.

This isn't to say that narrative doesn't matter; rather that the structure, systems, and mechanics, which all move in the background, are as much elements under inquiry and scrutiny. The game design document primarily unpacks the systems lying in the game background. Put simply, the document provides the rules of the game. It is a detailed breakdown of how the game's systems interrelate and how the user's actions impact those systems.

Throughout the construction of this text, I continually struggled with how the linear narrative of text offered the only opportunity to engage with the material I was presenting. There was only one way to go, and if readers were dissatisfied with my interpretations of the material, there was little they could do to shift the trajectory of the framework I had constructed for them. In short, *Developer's Dilemma* wasn't much of a game. Maybe it should have been a *Choose Your Own Adventure* book? If you agree with

Latour's frame, turn to page XX or if you think we should follow Smith's lead, turn to YY.

This was particularly troublesome because the framework I had constructed was more complex than I felt I could adequately convey. Not to mention that my informants had impressed upon me the importance of play. The field was continually shifting under my feet and demanded more methodological nimbleness than I felt my current form allowed. I turned to the very terms of my informants to find new discursive and technoscientific resources to deploy in my analytical toolkit.

I turned to concepts like the "vertical slice" (VS) as a tool for ethnographers to think through issues of contiguity, speed, and proximity that the field holds for those working in highly mediated contexts. The VS also gives ethnographers a means to think through the complex feedback loops between the observer and the observed. It is where informants read and respond to us, often when we are least equipped to deal with new information.

The iterative, or "interactive" approach to the field—our data, our methods, and ultimately our findings and subsequent ethnographic narratives—provides the opportunity to more reflectively engage with the field and our informants. Though it might be said that this differs from more traditional ethnographic practices, for the most part anthropologists of science and technology have acknowledged that ethnography must shift in order to account for "how various kinds of systems (textual, social, technological, etc.) hang together," and how those system "are continually being reconstituted through the interaction of many scales, variables, and forces" (Fortun 2006, 296). It makes sense that in this context our research projects become more complex.

But this isn't "anything goes." There are underlying systems and structures that I am attempting to grapple with in this. There are rules to the game being played. The difference, of course, is that I ask for readers to engage with them, likely making their own conclusions along the way. And, perhaps that isn't any different than the state of affairs in all texts.

After all, the process of design is rooted in constructing systems within which narrative systems can emerge. More important, perhaps, than any narrative I want to hammer into the minds of my readers, is the story that emerges for them from reading this text. As such, I attempt to lay bare the underlying systems that my instrumental play of this system has revealed. In the end, it is the demand that one's reader participates and co-constructs (or co-deconstructs) the narrative that makes the difference. It is the demand that a text or argument be played that assumes a different kind of engagement assumed in "reading."

Playing the text, of course, allows for hacks, mods, remixes, and cheating, much like gaming in real life. It is perhaps for these reasons that I pursue the game-form as a supplement to the ethnographic text. My readers deserve an opportunity to play the field as I did during my years of participant observation. Can their forays into my virtual space ever be equivalent to the fieldwork I pursued? No, but that isn't really the point. Perhaps the points are multiple.

My hope is that the interactive/engaged instrumental play is an opportunity to engage with the structures as I perceived them and either validate my findings, or make them swerve, though likely both will occur.

The opportunity to play the story resonates differently, much like the game developers making games, or students playing games—the process of the push and pull allows for alternative readings. The game of tug-of-war that we play between systems that restrict as we attempt to muddle through them is an opportunity for my ethnographic accounts to have greater verisimilitude. They make sense and are means of making sense.

I make no promise that the game-form is ultimately superior or inferior. It is all in how things get presented. There are certainly better ethnographic game-forms than others. Each "text" illuminates and simultaneously obscures aspects of the overarching ethnographic narrative, one that ultimately emerges through play and replay. I continue to muddle my way toward both an ethnography and an ethnographic game that makes sense of the worlds of videogame developers in the context of global neo-liberal capitalism.

The explicit engagement with design as shaping the resulting possibilities for collaboration, interpretation, and remixing encourages attentiveness to the construction of the ethnographic argument. The game-form ultimately offers anthropologists new means to approach their objects of concern as well as new collaborative opportunities for readers and informants.

What I want to stress is that the game-form offers both promise and peril for ethnographer and informant. More often we need our ethnographic forms to take new shapes. This is not to say that we need to jettison one type of text in favor of the other, simply that we need to be thinking more broadly about what constitutes the ethnographic narrative and what possibilities can be found in new forms.

World 8-2: Vertical Slice—An Analytic Conclusion

This final world returns to the central category of the text—creative collaborative practice—that resides at the core of this text's analytic focus. Analyzing creative collaborative practice in the context of videogame development requires analytic attention at different scales. The entire system is important, and the text grapples with this system. World 8 is one way I have attempted to wrestle with how social analysts can come to confront systems that cross "scales, variables, and forces" (Fortun 2006, 296) in different ways. The game as an analytic tool is one that I investigate as a means for understanding the complexity of these formations as well as how they are open to change and (re)interpretation.

What kind of game would the game industry be?[1] Would it be a fun game? World 8 mobilizes the arguments made throughout this text and synthesizes them by conceptualizing the videogame industry as a game——a designed game. It is an opportunity to ask different questions about the work of videogame development. As this text has conceptualized it, the game industry is a multiplayer game, though not necessarily a "networked" game in the traditional sense like a first-person shooting game. It is both collective and individual. The individual affects and is affected by the collective. At the level of work practice, this exercise makes the point that interactivity, though a valuable tool for game developers, can also be pushed too far. People must be given the time and space to get work done. Crunch is, in many respects, the product of over-interactivity in concert with poor planning, modified timelines, artificial demands by other interests, and the continued demands for secrecy in the game industry. To understand the interconnections between these aspects coming together in concert to produce crunch as it now exists in the game industry prompts a different kind of text: a game design document.

A game design document for videogame development work embodies the foundations of a kind of procedural rhetoric, and this text functions as just such a document. The game industry game is "persuasion through rule-based representation and interactions rather than the spoken word" (Bogost 2007, ix). This design is an argument about the structures that shape and are shaped by videogame developers. The advantage is that it is open to a multiplicity of interpretations. Though the industry game is not yet a game that can be played, it is nonetheless quite serious in that I have attempted to think through the kinds of structures and play that would occur in this space. It is also a method by which to think through a different form of ethnography.

The account given in the preceding chapters is one possible story that a cultural analyst could draw from the material gathered throughout my fieldwork. Although that "account(s) may be truthful; . . . [it is] in principle, susceptible to refutation, assuming access to the same pool of cultural facts." In other words, this is not "*the* story, but a story among other stories" (Clifford 1986, 109). World 8's ethnographic game is an attempt to create a persuasive yet playful space where alternate interpretations or arguments about the structure of *Developer's Dilemma* can emerge. In this way, the ethnographic game takes an entirely different view of the ethnographic allegory, which attempts to limit the play within the text. Instead, the ethnographic game invites play. Of course in some researchers' eyes this will discount the scientific character of this study. Yet I strive for that goal of many game developers with too little time and too few resources who attempt to capture a sense of reality—verisimilitude.

This game, as imagined here, is designed for the Nintendo DS for numerous reasons. The first is personal: the Nintendo DS has a burgeoning homebrew and technically illegal community growing around it that I hope to support by offering new arguments in support of their activities as well as the technical resources created during the eventual development of the game. It is my opinion that developing a game for the Nintendo DS in connection with a scholarly project demonstrates the illogical character of criminalizing speech on proprietary technologies. The ability to speak with and through devices that are owned by a user should not be compromised by legislation encouraged by corporations that largely have been unable to prove the value or sustainability of what they promote.

The other reasons the DS is the target platform for this industry game are more practical. More than any other console device, the DS is the closest developers and gamers have to a universal language in that both hardcore gamers and casual gamers alike have accepted the DS. The DS has a much broader player demographic than typical console hardware. Nintendo has gone as far as marketing the device and a subset of its games at numerous markets—men, women, young, and old. Atypical games have been released for the console and it continues to attract new kinds of gamers.

Another reason for creating this text as a DS game lies in my findings from those who work in the industry. Thinking broadly about who could conceivably be or desire to be game developers is an important aspect to changing the structure of their work communities. In some respects, the decision to target the DS is a departure from using this exercise strictly as a demonstration of the structures of the game industry. If that were the goal, then I would have targeted only the PlayStation 3, the most inaccessible and expensive of the current generation of consoles.

The DS's two screens, stylus input, built-in Wi-Fi, and relatively low price also make it an attractive target for this project. The two screens allow the simultaneous interaction and presentation of new information. The stylus, while being an approachable form of input for gameplay, is also more like the mouse on a personal computer, where most games are actually created. The DS's Wi-Fi capabilities allow "teams" of developers to work together on tasks. I do not conceptualize this collaboration in "real-time," but rather through a mechanism where individual tasks come to affect others on your team. Each individual developer has their own drives and specialties, which ultimately affects the kinds of games they play, as well as their progression through the game. The graphics are meant to be stylized and again, accessible. Rather than appealing to core demographics, the DS emphasizes accessibility.

Over the course of gameplay, developers hope that players begin to understand their position in a larger structure, some characteristics of which they can adjust and modify, and others they cannot. It is these structures I hope players will come to question and desire to change. The game's design is meant to bring a broader understanding of what it means to create videogames, as well as appreciate the work that numerous people do to bring games to market, especially since most are unacknowledged or never seen as integral to the development of this class of new media.

Each world of *Developer's Dilemma* has introduced an analytic category, which I found productive in the analysis of the everyday lives of game developers. World 1, via its structure begins to introduce and explore the everyday lives of game developers. This is important because the text itself embodies a crucial aspect of the videogame industry and work of game development: game worlds are rife with contextual information that needs to be examined closely. Too often games and game developers are not granted close readings. By structuring the text as a tip-of-the-hat to *Super Mario Bros.* I acknowledge my ancestry as well as that of so many of my informants. The very text and structure of *Developer's Dilemma* attempts to bring some of the contextual systems of game developers into view and decode them for the reader.

The frame that mobilizes World 2 encompasses underlying systems and structures and the drive that game developers have to find them or construct them. This framework uses the concept of instrumental work/play as the mode of activity through which developers arrive at the kinds of conclusions made day-in and day-out. The discussion of instrumentality in World 2 is about discretely and precisely identifying each element of a game that has yet to be constructed; it is both world building and world deconstructing, simultaneously. In some cases it thus appears that game developers are keeping others at bay through their culture and language, yet this is not simply done as a means to maintain exclusivity, but rooted in a broader attempt to define a kind of epistemic space of their own.

World 3 draws on the notion of the experimental system as a means for understanding what game developers attempt to do in their preproduction work. The experimental system of the game is not just the game itself, but the systems and people around it that support that system. Thus, the experimental system isn't simply an apparatus, but comprises the supporting humans and non-humans, often drawing on different world views and the means by which they come together in fault line-ridden ways. Yet, this is precisely what makes games particularly interesting systems for developers to construct, rooted, as many of them are, in gaming and gamer history. The riskiness of game development is precisely that this experimental system may very well fail. Time and again, the juxtaposition of these diverse backgrounds, systems, and epistemic perspectives keeps game developers passionate about their work.

As we transition from the temporal space of preproduction to production in World 4, game development takes on a very different character. It is here that experimental systems are expected to function flawlessly in an interactive way that keeps developers collaborating and moving forward. All of the meticulously designed (or not so) practices, technologies, and approaches place strain on the experimental system. Things begin to break down and the hope or promise of interactivity gives way to something more problematic; developers begin to lose perspective on the overall whole of the game. Individual tasks, the build, and "whose piece of the puzzle isn't meshing into the overall whole" become the focus. Interactivity becomes a goal, rather than an element that enables the creative collaborative practice of game development. World 4 contextualizes developers among the myriad of tools they use to construct their games as well highlighting the dysfunction of getting a game to function.

World 5 descends into the moments where the wheels simply start coming off the cart. Interactivity gives way to Autoplay, or complete engaged disengagement. Crunch envelops teams attempting to meet deadlines while changes are dictated from afar. Systems and practices begin to fall apart under the stress and strain. Yet, the perpetual startup system requires this, and the difficulty of breaking into the industry combined with a ready stream of talented, passionate, and young people—enable it. And all of this is rooted in the underlying systems and structures of the broader game industry, which is why we progress to the next level.

In World 6 we see explicitly how access, control, and secrecy go hand in hand with the numerous issues that developers face. The actor-network can be structured and disciplined in different ways. Access to particular technological or social obligatory passage points (such as the DevKit, NDA, and licensing) maintains certain kinds of actor-network structures. These elements are part of what lead to our perpetual startup cycle, which, viewed from another perspective results in institutional Alzheimer's. The evidence would indicate that this is not a healthy direction for the game industry. While publishers and manufacturers talk in directions that would seem to enable new kinds of institutional memory, in reality these claims are merely words said to appease earnings analysts, rather than actions that result in improvements for the everyday lives of game developers who in the end are responsible for the creation of the games these organizations depend upon. Ultimately, it is the everyday working game developers who are responsible for creating games and the happier, more sustainable, and diverse the community of game developers, the more likely it is that developers will remain in the industry. Experienced, rested and happy developers can only improve the products of the game industry.

In some cases, the active disciplining of actor-networks is so severe and so counter to the claims of console manufacturers that it deserves specific attention of its own. World 7 turns to the ways in which software, hardware, and the ability to mobilize governments to seize property and imprison individuals further entrenches the existing structure of the game industry. Despite the fact that many of these companies now recognize that this particular state of affairs is not working, and that the rise of independent game development indicates those old structures do not maintain quality and innovation as much as they were once thought to, the industry nevertheless continues to co-opt State powers to maintain control. Those very systems that publishers and manufacturers cling to so tightly, however, may be the very vehicles of their own destruction. Of course in the meantime, many game development studios may pay the price.

Finally, in an attempt to bring the text full circle, World 8 returns to the structure of the game and a game design document. The final level of a game should agglomerate each and every element that the player has been expected to master and ask players to leverage these skills to prove a kind of mastery over the entire system. What better way to consolidate our understanding of designing games than to think through what kind of a game this might be? The contribution of World 8 is the realization of my argument that game developers have a very particular epistemic perspective that ought to be respected as such. A game is just as capable of conveying complex frameworks as a conventional text, and does so in such a way that might just compel you to keep playing, despite knowing that it isn't entirely working quite right.

World 8-3: "The Expo Demo"—Two Gameplay Narratives

What follows are two gameplay narratives as imagined by two hypothetical players of the game *Developer's Dilemma*. It is thought of as a conceptual or verbal account of what a "playable demo" of the players of the game would experience.

Narrative 1

The first thing I was presented with while playing this game was a screen, which I was asked to enter my name. I presumed I could enter whatever I like, but I chose to enter "Cassie O'Donnell." The second prompt asked me what my college undergraduate major was. There were several options, I selected "computer science." It told me that I was currently unemployed, had managed to escape my undergraduate institution with no debt, but that I was "unfortunately" located in the Midwest, where few game jobs were available.

I was presented with my developer's "status" screen. It indicated that I was currently unemployed and I had "undergraduate" skills in engineering and "low" skills in art, design, and management. My personal status was currently "happy." When I touched that element with my stylus it went into more detail, displaying that my "fatigue" was "low," my mood was "good," number of hours at work was zero, I had $1,000 in the bank, I was single, and I had no children.

Four new options presented themselves: "search for jobs," "create independent studio," "join independent studio," and "relocate." I first investigated the "relocate" option, because I had previously been told that my location was "unfortunate." More unfortunately, the $1,000 I had banked was not enough to finance relocation to anywhere with greater game development job availability. So, with that in mind I selected "search for jobs" and began searching for a job. There were several available "software development" jobs for which I applied. I was accepted and took on the role, "entry level software engineer."

At this point I was presented with my first engineering puzzle game. It was relatively easy to complete. My goal was to trace the movement of "data" into a system and correct "improper" data movement. Several pieces of data were being improperly placed, which I corrected by adjusting "pipes" on the screen. After completing the first game I was returned to my "status" screen. My money was gradually increasing and my engineering skills were increasing. The avatar's mood was happy and well rested, though less rested than prior to taking the job.

I then chose to "create independent studio." I named my studio "Alchemyst Creations." I entered into a new engineering puzzle game. During this game however, I received a message that my "real" job was demanding my time. I could either respond to the request or continue with my independent work. At first I selected to continue with my independent work, and was then instructed that this would likely result in a poor performance report from my real job. Considering this warning I instead chose to return to my real job. I completed two more engineering puzzles before I was returned to my independent engineering puzzle. After completing that puzzle I closed my console for a break.

Later, when I returned to the console, I was told that I had two waiting real job puzzles to complete. After finishing those I began another independent puzzle. My independent studio status screen indicated that "production" on my first game was 5 percent done with respect to engineering, but 0 percent for both "design" and "art." There appeared to be several available options: to "take on design task," "take on art task," or "find other developers." I tried the latter option, at which time the console attempted to wirelessly find other "developers" in its range. I was by myself however, so it could not find anyone else. It encouraged me to find players with artistic or design skills and "connect" with them using this feature. It also said that I could "manually" add other players using the "friend code" option.

In the meantime I took on several art and design tasks. The first art task was to duplicate several line drawings displayed on one screen using the stylus on the lower screen. Other tasks involved attempting to place a texture on a 3D model by selecting points on the lower screen. Many of these puzzles were quite difficult for me. Later that night I emailed a friend from high school, asking if he had seen this game. He had not, but downloaded it and placed it on his console. Since he was an artist he was initially assigned a job at an advertising company, but was able to join my independent studio using the "friend code" option. At this point we both became able to work together on our independent project, when not assigned tasks from our real jobs. He took on the art projects, I did the engineering, and we split the design tasks.

During this time, my position at my real job role had improved to "lead software engineer," and I had begun being assigned "management" puzzles. While my management skills were characterized as "improving," the work made it more difficult to improve my engineering skills, which were crucial to my independent project work. My avatar's mood had shifted to borderline unhappiness and the number of hours being worked had steadily increased. I began to seriously consider the relocation option available in the game. It was going to cost me a significant amount of money, but I could relocate to the West Coast where there were several available "jobs" at established game companies. I found that I could search jobs in these locations and even apply for them, though I frequently only received rejections or no response at all.

Eventually I did decide to relocate within the game. At this time I was able to take an "entry level game engineer" position at a company for less money than I had previously been making, but it was an opportunity to be a "real" game developer. Very quickly I was being assigned new puzzle tasks. On several occasions when I returned to the console after having shut it for a day I would find myself with five to ten puzzles to complete before I could return to my independent work with my friend. Then one day the game notified me that game production at the company I was working for was entering "crunch" mode.

I wasn't quite sure what that indicated, but it initially meant that I had puzzles coming at me nonstop. When I attempted to close the console, it flashed a red light, which I assume indicated something, so I opened the console again. It said that if I chose to stop in the middle of this puzzle I would be risking my job, and that I should finish this puzzle and two more before closing the device. I did that, reluctantly, and closed the console. When I returned to the device the next day I had fifteen puzzles waiting for me to complete. Quickly I found myself working exclusively on these puzzles. Occasionally other puzzles would interrupt the puzzle I was working on, and not even reduce the number of puzzles I needed to complete before closing the device without risking a poor performance review.

I noticed that my avatar's status was deteriorating. This began to manifest during puzzle activities, where "bugs" or "mistakes" would strike while I was attempting to solve a puzzle. These would frequently make the puzzles more difficult and take more time. The character's fatigue was increasing and the mood rapidly turning toward very unhappy. Finally I was frustrated enough with this process that I began searching for new jobs. There were plenty of new jobs available, so I tried a new company. I was even hired as a "senior game engineer," but quickly this company too was in crunch mode. However, I had saved enough money up that I quit my job this time and began working exclusively on the independent project.

Unfortunately my friend's avatar was not doing much work on the independent project by this time, because he was working for a game company as well. However, I knew a handful of people with the game now, so they joined my company as well and began working on the project. When the game was 50 percent complete a new option became available. We suddenly had the opportunity to "shop your game around" to publishing companies. When we did find a publisher willing to fund the remaining development of the game, suddenly new tasks began presenting themselves, primarily management tasks. Because I had started the studio, everyone indicated that I should handle those tasks.

Our independent work had quickly become our own work, the publishing company began also asking for changes to our game, and new engineering, art, and design tasks began presenting themselves as a result. In an effort to meet a deadline for the game (part of the deal with the publishing company), I was forced to indicate "crunch" mode for our game. It became readily apparent that this was the same game all over again, only I was in charge of the company this time. Some of my fellow players began to bail out, resulting in more tasks for fewer people. I tried to bring on other players, but they too quickly dropped out of the game. Eventually it was just a handful of hardcore players that made sure that the game completed.

While there was a sizable payout at the end of that part of the game, my avatar was left "exhausted," and almost incapable of completing a puzzle due to the frequency of bugs and mistakes. I set the game aside for a while to recuperate myself as well. At some point it seems inevitable that I will have to use the "leave the game industry" button, an option always available. Leaving seems such a shame after all that I have invested.

Narrative 2

As I sat down to play the game, the first question it asked me for was my name. I entered "Maria Murali." It then prompted me for my major in college. I selected "fine arts," as it was the closest thing I could find. It indicated that I was located in central India, which was indicated as "fortunate." I wasn't entirely sure what that meant, but I assumed it was a lucky break.

The game's next screen indicated that I was unemployed with "undergraduate" skills in "art" and "low" skills in engineering, design, and management. My avatar had a smile on its face, which I assumed meant that I was happy. I later discovered that by tapping on the avatar with the pen that it showed a more detailed breakdown of what indicated my happiness. I had spent 0 hours at work thus far and had only Rs. 1,000 (approximately $25). I was not in a relationship and had no children.

The game presented several available actions at this point. Of those, I selected "search for jobs." There were twelve available "art production" jobs available in my vicinity. I applied and was accepted as "entry level production artist." From this point forward the game spiraled into a series of different art-based puzzle games. Goals ranged from things like simplifying basic 3D models to use fewer polygons to filling in the proper colors on a mesh to meet a goal model's appearance. One of my favorites was attempting to reproduce a higher-resolution image as a much smaller piece of pixel art.

Throughout this process I could see via my avatar's status page that my artistic skills were increasing and my money was increasing as well. I guess that meant that things were going well. However, my avatar was getting tired, so after finishing several puzzles, I closed the DS and walked away from it. A few minutes later I walked by and observed that the DS was flashing one of its indicator lights at me. I opened the DS and the game informed me that there was further work to do. This struck me as strange, but I completed a few more puzzles and closed the DS again, observing that my avatar was quite tired.

Despite the fact that I saw the indicator light flashing a couple of hours later, I ignored the DS for a day, the game not being a priority at the moment. When I next opened the DS, it informed me that I had been "let go" by the game company that had hired me and that my status was "unemployed," I did observe, however that my character was happy and had more money than it used to.

Again I selected "search for jobs." There were a number of "art production" jobs available. I scanned through the list and selected another. Again, the game launched into a series of puzzles. I would play for a bit, set it down, and come back. Eventually the game would indicate that my employer was unsatisfied with my work and I would be "let go." It was a bit strange, but there were plenty of jobs to be had, so I didn't mind the process.

After several weeks, an "art production lead" position became available. I applied and was granted this job. It included a new set of puzzles that were about managing other artists in addition to doing art-based puzzles. Occasionally, management tasks would override an art task, but overall the process was still interesting. My character would get fatigued, but I still managed to get enough puzzles done at each sitting that I kept that job for quite a while.

One thing struck me as strange during some of the management processes. Occasionally, I had the feeling that my success in puzzles resulted in poor employee performance or even employee loss. The game did not penalize me for this, but I was curious about the process. It was also during this time that I observed errors in the game. At first I thought they were bugs in the game itself, but it became clear that some puzzles were actively attempting to thwart my efforts at solving them. I noticed that this tended to only occur when my avatar was particularly tired.

I'm not entirely sure when the game became a compulsion, but I played it a lot. Despite my character's obvious fatigue and the active negative feedback the game gave me, I played a great deal. My character's skill continued to climb in art and management and so did my bankroll. It was about this time that my parents introduced me to my now husband. I set aside the game and never really came back to it, though I've thought about it on several occasions. I never really had a chance to see where I could go, or what opportunities would come about. The thing that concerns me the most was that compulsion I felt to respond to its indicator light that there was work do be done and I was needed.

World 8-4: Core Gameplay and Game Elements

The work of game development functions as the core gameplay mechanic. It is stylized in the form of puzzle-like tasks that the player navigates. These tasks attempt to approximate the "play" of work as much as possible, though basing work in play does not undermine the fact that the puzzles can be difficult or complex. These tasks take the form of "mini-games" that the user plays. The overall goal of the game is open ended. Players can determine if their goal is to "create titles," create their own company, climb the corporate ladder, or simply enjoy the tasks which they have the opportunity to work on. While there is no "score" in the traditional sense, several different sub-systems have the potential for being recognized as a "score." Each of the employment history, employer, and titles published subsystems serves as sites of potential value that players might view as being a score. The game has several underlying subsystems that each impact the overall game mechanic. These are divided into the categories enumerated in table 8.1.

Some of these sub-systems are simpler than others. "Employment History and Titles Published" are a historical record of what a player has done through the course of the game. Some are more complex. The "skill level" system is one of the primary game mechanics, as it determines the kinds of employers and job roles available to a player. Players are awarded "experience" during the course of playing mini-games. Games conceptualized as "engineer" games will be heavier on engineering skill rewards. The player then has the ability to distribute these points among those skills (listed in table 8.2) they would like to improve.

Table 8.1
Game subsystems

Subsystem name	Description
Skill levels	The skill level represents skills acquired while working. These skills provide further options during the play of mini-games as well as the ability to advance or move to other companies.
Personal status	Personal status represents the state of a worker. Fatigue, mood, number of hours at work, money, relationship status, and number of children are included in this category.
Regional location	The regional location is randomly assigned to the player at the beginning of the game. The region affects available job opportunities and employers. A player can move if they have enough money to do so.
Employment status	Employment status represents where a player is employed.
Job role	Job role is the kinds of primary tasks available to and assigned to a player. Sometimes jobs not associated typically with a role will be available, providing players with the ability to gain new kinds of skills.
Employer	The employer of a player determines the kinds of tasks available to them. Players can create new "independent" companies while employed at other companies, though they must maintain their work level enough to remain employed.
Employment history	The employment history of a player is a record of where they have worked, how long, and kind of work they did.
Titles published	The titles published are a list of games that the player has been credited for in their work.

Each player, when starting the game, will choose an undergraduate major. These will affect the starting values of a player's skills. All players will start out unemployed, and can either begin searching for work, move if money provides the opportunity, or start their own game companies. These companies of course will only have one role available to begin, "jack of all trades." Numerous tasks will be available to the player, but they will quickly have to decide on a specialization and then either pursue employment at an established company, or attempt to attract other players to their companies. At any given time a player can only be part of two different companies. This amounts to the idea of a "day job" and a "startup." Of course a player can belong to two startup companies, though this will quickly affect fatigue, mood, and money.

Table 8.2
Skill-level subsystem detail

Skill levels	Category and description
Software coding	[Engineering] Coding skills point at the ability for a player to quickly complete software development tasks. This is not necessarily indicative of the quality of the code being produced.
Software design	[Engineering] This is the other side of software development skills: being able to produce "good" or "well designed" code.
Debugging	[Engineering] Debugging is the ability of an engineer to determine where something is going wrong in the process of game development.
Modeling	[Art] Modeling is the creation of objects that can be placed into the game.
Texturing	[Art] Texturing is the creation of skins for models.
Animation	[Art] Animation is the ability to create the animation sequences that textured models depend upon to be put into motion.
Level design	[Design] Level design is the creation of game environments.
Character design	[Design] Character design is the generation of compelling game character concepts.
Game mechanics	[Design] Game mechanics are those rules and systems that underlie a game's visual presentation.
Scheduling	[Management] Scheduling is the ability to accurately estimate and hit project deadlines.
Resource allocation	[Management] Resource management is the ability to keep a team on track and adequately tasked for a project.
Networking	[Management] Networking keeps developers connected to other developers, publishers, and manufacturers.

Some game systems provide restrictions on the player. Personal status, regional location, current job role, employment history, and titles published fit well into these categories. The personal status of a player affects the amount of time they can reasonably remain at work each week without their fatigue increasing or their mood falling. These two categories will affect the "accuracy" with which they can complete work-related tasks. The more fatigued a player is, or the worse his or her mood is, the more possible it becomes that actions taken in a task will not work quite as desired. This is done to approximate the declining ability to remain focused experienced by fatigued workers.

The "game is over" when the players mood sinks to "jaded" and their fatigue reaches "hospitalization," or at any point their money reaches zero. This is considered "burnout." A player can later reenter the industry with a character if sufficient real-time has passed since the game was last played. Players can also opt to "leave the industry," at any time during the game, though players who are working with that player will suffer in-game repercussions.

In the game that is the videogame industry, the regional location affects the likelihood that particular kinds of employment are available for a would-be game developer. Each employer defined in the game will have regional locations where particular job roles are available to the player. These restrictions will reflect the available job opportunities of a given geographic region. Players will be equally likely to spawn in the United States, Western Europe, India, or Japan. In locations where game development jobs are particularly rare, developers will have the opportunity to work at companies that border game development work. Engineers will have the opportunity to work for software companies, though they must maintain a company of their own on the side to "remain in the industry."

The mini-games represent the game space of the videogame studio. Mini-games are tasks which each player must complete. As a player moves into certain kinds of positions or organizations, tasks may become timed. An "estimated" time may be provided, which a player must beat in order to gain all points associated with a mini-game. Timed mini-games may not be paused or stopped by closing the DS without penalty. It is assumed that if a timed task is paused that the human player needs a break that would be unavailable to a worker. While the player may pause, their virtual character's skill status will decrease during these situations, correlating with a drop in professional status as one avoids assigned work. Untimed tasks may be paused or the DS can be closed and put into its hibernation mode.

Crunch mode occurs in the game when a series of tasks are scheduled back to back, and the user has no option to not complete all of them simultaneously. Again, if the game is paused or place into hibernation, the characters' skill status will be decreased. If crunch has not occurred recently, a temporary increase in mood will be provided, representing some of the allowances made to employees during crunch times.

Throughout the play of mini-games, "interactive" prompts from other employees as well as other games may interrupt the flow of a current game. Meeting mini-games, comments from fellow workers, email messages, and instant messages will in some cases distract a player. If a player has a significant other or children, then additional mini-games or interruptions may occur, resulting in fewer skill-level experience points. Over time this loss of points may result in lower earnings or fewer advancement opportunities.

Players may also be dependent upon the completion of a task by another player. They then have the opportunity to send an IM to that player to check on the progress of the work. These messages will later show up on the screen of the other player in the form of "interactive" prompts. However, these same prompts can distract or affect the work of the other player being done. They are designed to be a double-edged sword.

Based upon the "amount of time spent at work," effectiveness at work, and other minimum values, players may be encouraged to leave a company. In many cases however, these players will likely "burn out" prior to a prompted end of employment.

Other game elements are defined by the employer and role of a player. In some cases, the combination of particular employers, roles, and locations may result in situations where players are pushed more quickly toward burnout. More sustainable models may be in place with other companies, roles, and locations. When the game is first developed there will be several categories of employers, though the list found in table 8.3 can be expanded in future downloadable updates.

Once a studio has reached the level of third-party or higher, they can begin "researching" tools to assist in the development of their games. These tools must be individually researched. If a player changes companies, these tools will not move with the player. If a player transitions from one company to another, they may gain access to new tools that assist in their work.

It is also possible, however, that during gameplay, if a tool is used it may malfunction depending on how frequently it is used by employees. The more frequently a tool is used, the more reliable it becomes. The less used, the less reliable. This means when a tool is first created it may actually make work more difficult. Over time however, and with continued use the tool will begin to simplify the work processes. Because tools cannot be shared, the testing of a tool must be done at each individual company, so players may have to experience the "learning curve" or "testing phase" of a new tool several times during the course of their career. Tools may also be purchased from a middleware company, while the cost of purchasing a tool may be less expensive than researching a tool, the learning curve and testing phases will still be required.

Tool access can also be restricted by geographic location and employer type. Some middle-ware will simply be too expensive for companies to purchase, while others are unattainable because employers have not yet gained access the those networks that provide them the opportunity to acquire tools or even know that their development is useful.

Table 8.3
Employer categories

Employer type	Description
Independent studio	Players can create an independent studio of their own at any time. They will either have to complete all of the tasks associated with game development, or they will have to look for other players to join their company. As these companies mature, they may be approached by publishing companies or manufacturers to do work, at which time the classification of their studio will change. In many cases moving from independent to contract work will be a necessary step for getting a job with any of the other available employers. These studios may be sold, bought, or merged with other studios, publishers, or manufacturers.
Third-party studio	Several third-party companies will be created for the game to serve as starting locations for players to work. These will be restricted by geographic location.
First-party studio	These studios frequently have more freedoms than third-party studios as they have more funding coming from a publishing or manufacturing company. However, these companies can also exert force over the studio during the development of a game.
Publisher	These companies will likely only be accessible to developers with high skill sets. Players will have to ensure that they live in the geographic locations where these companies are, and that their skills are such that they will be hired.
Manufacturer	Similar to publishing companies, but a step higher in difficulty for developers to gain access.
Software firm	In locations where engineering game development work may be more difficult to find, players interested in being engineers will likely have to work for a software firm while also doing independent game development.
Art production studio	These companies only employ artists and typically have fewer "crossover" tasks that allow developers to work on projects that expand their skill set. However, these employers will train artists and pay them enough so that players have mobility if they wish to change geographic region.

Table 8.4
Mini-game categories

Game category	Description
Engineering	Engineering tasks involve an assortment of parsing and number crunching tasks. There are also debugging tasks, where the player must identify previously parsed or number-crunched pieces of work that are incorrect.
Art	Art tasks require the player to create and modify different kinds of artistic elements. This can be the creation of wireframe models that approximate goal images. It can be the picking of proper texture coordinates to skin a model, or the creation of animations for a model.
Design	Design games require the user to take game content and mechanics and solve problems with them. The goal is to make a particular goal event occur through the modification of on-screen objects.
Management	Management work is frequently puzzle work, piecing together disparate pieces of a system. The goal is project completion, or task completion, or the movement of resources from one location to another.

Each mini-game will be divided into one of the four categories listed in table 8.4.

Different mini-games can be defined for each category. Throughout the game, frequently players will have to engage with some games that fall outside their realm of expertise, or they may take a job that requires them to complete tasks outside of their area of expertise. In these cases, games may not behave entirely as expected, in order to simulate the accumulated expertise over time. However, if a player continually plays and does well with particular types of games and expands their skill levels at these games, new jobs, companies, and tasks will become available to them.

While particular mini-games will be encountered more than once, they will also have elements that are generated randomly when a game starts. For this reason, players will not be solving the same game each time. Rather, when they encounter the same game, it will begin in a random fashion. Engineering number crunching games will be randomly seeded in such a way that they remain re-playable. All games will have one or more randomized elements that will it possible to replay games.

Because players will also be presented with occasional games outside their area of expertise, it will give each player the opportunity to play different kinds of games. Players can increase their skills in new areas in hopes of receiving new kinds of games. New “moves” will be provided for players as their experience increases. For example, an experienced artist working on a particularly difficult model can push a button that says "basic layout" that moderately decreases the complexity of the model they are working on. In some cases companies will have defined special moves that assist their employees in getting tasks completed.

"Winning" is a category that players can define on their own. Because the game is designed to provide numerous user-defined goals, winning can be successfully running one's own independent studio, working at an established company, or starting a company that is acquired and leaving the industry with a large bank account, etc. The definition of winning is intentionally left ambiguous to encourage players to reflect on their motivations and goals within this open-ended system.[2]

World 8 Boss Fight: The Credits Roll and Bowser Lives

A game design document doesn't mean much. In the words of Erin Hoffman, the unveiled ea_spouse turned game designer, consultant, and game industry activist, "Your game idea actually sucks" (Hoffman 2009). What this design document does is return to the core systems that dominate the videogame industry. It is an opportunity to think through those structures in an alternative format that allows for divergent interpretations of the material. And so, perhaps my game idea actually sucks, but it is a game based on extensive data gathered from within the game industry. It is a game exploring life realities rife with systems and structures that are rarely talked about outside the yearly rant sessions at GDC and the occasional blog post signaling that another experienced game developer has burned out and spun out of the game industry ecology.

More than an attempt to actually create a game, this structure provides a commentary on the state of the videogame industry as it currently stands in a way that allows engagement and interactivity in ways that a standard text wouldn't. The goals available in the game, one of which a player must adopt—either the creation of their own studios, console manufacturing companies, or the amassing of money—is precisely the narration that I would encourage developers to defy, to play differently. And perhaps seeing such limited options on a menu will incite them to do that.

It is my hope that this account also encourages discussion around certain aspects of the videogame industry, like the artificial construction of restrictions that prevent access from developers in the United States and other countries. I believe it also makes apparent that features like "crunch" and a culture of overtime are not only undesirable for the game industry, but are the product of a particular construction and imagination of the game industry. They are certainly not an inevitability. I also hope that this game document encourages developers to appreciate the value of experience and expertise, for both are crucial to the future success and stability of the videogame industry. The continued hemorrhaging of talented hardworking individuals and the lack of collaboration that results from an environment demanding secrecy are problematic for the industry's longevity and growth.

Perhaps more than any other aspect of the game industry, the tools of game development have changed since I first undertook this research. It is important to note that console development remains largely a dark art, characterized by custom asset pipelines through which a game's art assets are processed for use in the game's engine. Custom build scripts and compilers must be used and configured to work with software engineer's preferred integrated development environments (IDE). Custom scripting languages must be integrated to each of those systems. Various middleware software, which simplifies or implements aspects of a game (i.e., physics or audio) must be combined with the various other systems that allow the process to function. This entire process is laborious in ways and for reasons that are far reaching (O'Donnell 2011b).[3] New upstart production tools like Unity (Unity Technologies, 2005) and game developer's increasing interest in mobile platforms have pressured even the most dominant "big game" engines, like the Unreal Engine.

That isn't to say that tools haven't progressed; they most certainly have. Systems like Scratch (MIT Media Lab, 2006), which offer integrated editing systems, limited models, and integrated remixing are linked strictly to their online platform. The amount of customization that can be performed with tools like Gamestar Mechanic (E-Line Media, 2010), another online system like Scratch that offers game models for users, is still limited. Even now, software packages like GameMaker (YoYo Games, 1999), GameSalad Creator (GameSalad, 2009), and Unity—game engines targeted at similar demographics—do not offer the kinds of integrated and highly platform specific editors that one would expect given the age of the industry. Each engine relies heavily on external tools used to create the content seen in the game. These software packages also often bear the scarlet mark of "hobbyist" development tools, not those of "real" game developers. Further, with the exception of Unity, none of those systems can be used to author games on modern game consoles.

Perhaps more than any other single technology, Unity has altered the shape and terrain of game development tools. Built as a largely platform-agnostic set of tools, Unity's asset pipeline makes it particularly interesting as a game development tool. Pipelines, as I discuss in this text, are one of the most ambiguous and least discussed aspects of game development. By making the asset pipeline a central element of the engine, rather than an afterthought or add-on, Unity simplified one of the most difficult parts of game development, simply getting things into the game.

At the same time, Unity's subsequent development clearly demonstrates a new interest in the core game industry, which is more indicative of the Unreal Engine, rather than in the hobbyist and independent developers that they first courted. Recent additions to the engine indicate a growing interest from mainstream "AAA" game developers, including the largest players in the game industry. Nintendo's latest console, the Wii U, includes as part of their Development Kits (DevKits) a license for Unity Pro. Yet, the gatekeeping to those platforms remains largely in place.

In 2012, many game developers cite the crowd-funding site Kickstarter as a fundamental game changer in the world of game development. I think that overstates the importance of upfront money for making games. Kickstarter is certainly one of many changes that have shifted the balance of power, yet it owes its leverage to the other changes that have made crowd funding a viable option.

On March 2, 2011, at the Game Developers Conference in San Francisco, Satoru Iwata, the chief executive officer of Nintendo took the stage for his keynote discussing the changing world of game development and distribution. Most surprising about his presentation was that, despite a slide including the title-screen of *Angry Birds*, a runaway independent game title for Apple's iOS from an independent game development studio Rovio, he never mentioned Apple and iOS (the underlying operating system of the iPad/iPhone/iPod Touch). At precisely the same time, across the street, also at the Moscone Center, the late Steve Jobs announced Apple's iPad 2, which included a significantly faster graphics processor aimed specifically at bringing bigger games to their platform. In that convergence, something happened, but what? Nintendo lost credibility; the future of the industry was laid bare; a secret shook itself loose; something. Nintendo was blind and self-protective, and Apple was changing things in ways indicative of broader changes throughout the world of game development.

The biggest shift in the game industry has been the rise of digital distribution platforms, like Apple's iOS "App Store" and Valve's Steam system. These breaks from the old models of physical distribution and the availability of new platforms for games have truly altered the landscape of the game industry. While Apple has been constantly critiqued for the "walled garden" of the App Store, the game industry is quite used to walled gardens, and Apple's walls are much shorter than those found in the game industry. At the same moment that Nintendo was expecting independent game developers to pay $2,500 for a Nintendo DS development kit, Apple was charging only $99 for access to their developer network. Even large game publishing companies have started making games for these platforms, which were initially dominated by individuals or small teams of game developers. The results of which are indeterminate? Is there a need for spaces free of those vested interests?

The rise of independent game development has benefited from these growing distribution systems as well as providing a means for paying for games in new ways. Digital distribution was certainly an option in 2008 but it wasn't nearly as respectable for games, and convincing users to pay was much more difficult. Even independent game development has changed from the worlds of hobbyist game developers to professionals who have left the industry to strike out own their own because of the new distribution and financial channels.

New economic models for game development have emerged, but have yet to be proven long-term. "Freemium" games that are free to play but encourage the player to pay for additional game content, and "social" games that leverage social networks like Facebook generate revenue in ways wildly different from the standard console-game model. Perhaps more surprising and controversial than social games has been the rise and fall of "gamification" and an intense interest to draw on the kind of engagement that many games bring for more "productive" ends. Funding has shifted in important ways that go beyond consumer and advertising dollars, as public and private foundation and grant dollars pour into this shifting industry.

Noting all these funding changes, then, we can agree that focusing solely on Kickstarter has been a bit of a distraction. Questions that focus on whether this new funding mechanism has "broken the game industry" miss the point. Steam, the Apple App Store, and Google's Play system made ventures on Kickstarter possible by opening up the ways people think about games, play games, and pay for games. The game industry now finds itself in an extreme moment of transition.

As such, this is not a climatic boss fight wherein the enemy is slain and a clear and decisive victory is gained. Both life and creative collaborative practice are a bit too ambiguous for that. Rather, this boss fight is the culmination of one account of one aspect of videogame development practice. It is an examination of much of the center of the game industry. There are indeed other possible avenues that developers can travel in the creation of their collaborative works. How does a different path shift the underlying structures? There are numerous developers experimenting with modifying how games are created, marketed, and distributed. What do these new experimental game mechanics mean for the rest of the industry? How does it fit back together? Like a game, these experimental systems are unpredictable and their highly interconnected character makes deterministic outcomes difficult.

And finally, what implications can be taken from this game and brought to bear on other aspects of work and new media production? What can be learned from the everyday practices of videogame developers? A great deal, including new ways to talk about and examine structures of domination and control. This boss fight is about the tension that remains between game development practices in the context of the global game industry. Worlds 2 and 3 examine the highly experimental nature of game development practice, particularly during the early phases of game development. Worlds 4 and 5 examined how those systems put in place to support development practices that unfold during game production begin to strain under the weight of production practices and external demands. Rapidly, game development practice moves from experimental and playful to mad scrambling mayhem. Worlds 6 and 7 then draw connections from those local worlds of game developers to the broader context of the game industry. World 8 brought all of those together as a kind of game design document speaking to their interconnections.

When I say "tension," between development practice and the industry, what I really mean is that the game industry survives, even thrives, in spite of itself. Rank-and-file game developers bear the brunt of the labor that comes with a slew of cultural and political-economic demands. The culture of secrecy that dominates the game industry is both top-down—non-disclosure agreements, closed licensing structures, proprietary hardware and software—and bottom-up—"my idea is super secret and super awesome," "nobody has ever thought of a space-based real physics resource management game this game is going to be awesome," "if I talk about this, someone is going to steal it." Compared to other "industries," the game industry is surrounded by secrecy about the daily practices of what and how games get created. This has led to a general lack of respect for the complexity and creativity of what may be dismissed by many as a toy on a shelf. Games have emerged as an important cultural form, yet game development is still often imagined to be something it isn't and the game industry has itself to blame for this.

In other spaces of cultural production, there are numerous standard tools and practices that enable greater modes of creativity. People are not restricted from talking about what they do or how they do it. The game industry's connection to software development practice ought to indicate a thriving connection between game development and the "open source software" movement. While that is truer for independent game development, it is largely not the case for much of the game industry. Simply recall the story from World 6 of TinyXML, ported by numerous developers to the same platform because of the NDAs that rule the Nintendo DS. While game studios may draw on Open Source Software, they can often not contribute their changes back to the community, and thus each studio replicates the work of others. There are a growing number of exceptions to the rule of secrecy at blogs like #AltDevBlogADay (.com), yet the fact remains that frank discussion of day-to-day game development practice is difficult to find. Game developers have no sense of professional identity, because so many feel so disconnected from one another.

Finally, game developers are not vocal enough about the absurd behavior of the companies for which they work and the policies, lobbying, and government intervention they seek. This is your industry. You make the games. Speak up. Nintendo, Microsoft, and Sony cannot revoke all of your licenses. They depend on your work for their survival. Clearly, the emergence of Apple's iOS devices and App Store, based on a walled but more open garden clearly indicate that there are other approaches to licensing and distribution. Developers' selections of platforms to support are as much a vote for the kinds of policies that they agree with. Vote with your labor. Go make great games.

Notes

World 1

1. I recorded most of my field notes digitally, in text files on my laptop. The majority of these files were created with TextWrangler, a freely available text editor for Mac OS X. I hand coded my notes, and inserted them into the bibliographic software Bookends that I used as both a means of data storage and analysis in grouping coded entries together. Interviews were also recorded digitally on an iPod via a Griffin Microphone adapter. I transcribed my interviews using ExpressScribe for Mac OS X and a USB transcription foot-pedal. Interviews were then coded using TextWrangler and entered into Bookends similarly to my field notes. All of these files were consolidated on my computer and encrypted using Apple's Mac OS X FileVault functionality, which encrypted the data using the AES-128bit encryption algorithm. Four primary codes emerged from these activities, "work/play," "interactivity," "networks of access," and "corporatization of the state." Each of these primary categories was also broken down into subcodes. I coded and recoded the existing literature, interviews, and field notes in Bookends. Using the "Smartgroup" categories in Bookends, I was able to create categories that automatically updated themselves with elements with particular codes—"work/play AND secrecy" for example. By selecting a Smartgroup, I was able to see all the material I had associated with my codes. This was used to organize the material contained in the text.

2. Later VV was contracted to produce a version of the game for Nintendo's Game Boy Advance (GBA) system.

3. I have written at length how this complex interplay of interests, combined with digital format "lock-in" makes many of the promises of digital "convergence" seem to disappear. These issues are critical in future discussions surrounding "transmedia," though the realities of what that means for content producers has been under explored (O'Donnell, 2011).

4. This particular aspect of globalization is one on which I presented and published early in my graduate academic career, looking at *Wired Magazine*'s construction of

Free/Libre and Open Source Software (FLOSS) and Indian new economy laborers (O'Donnell 2004a).

5. This material is based on work supported by the National Science Foundation under Grant No. 0620903. Any opinions, findings and conclusions or recommendations expressed in this material are those of the author and do not necessarily reflect the views of the National Science Foundation (NSF). That research was conducted under the guidelines of Rensselaer Polytechnic Institute's Institutional Review Board (IRB) approval #723.

6. A failing of this text, as I see it, is a lack of insight into why, despite a changing gender demographic among the players of games (Datamonitor 2004; Datamonitor 2005; ESA 2005; Datamonitor 2007), the demographics of game developers have remained relatively static and continue to be significantly lower than those who play games (Gourdin 2005). This failing is partially my own. A lack of women informants at primary field sites is a cop-out. Those informants are certainly present in the text, but I should have pressed the issue. The majority of the US studios visited as part of this research exhibited demographics similar to those of the rest of the country. As such, women developers are clearly included in this account, but most did not reflect on their gendered position. Many women developers talked about it as something they had become used to, in the process of pursuing game development work. In other cases, they were actively uninterested in talking about it, for it marked their gender in ways they were uncomfortable with. Perhaps I focused too myopically on specific corporate sites throughout the research. I think it is critical that future research of women game developers should be undertaken, informed by this work, to better understand what compels and constrains their progress in the industry. Certainly, recent controversies, like those on Twitter surrounding #1reasonwhy, #1reasontobe or the vitriolic comments that appear when women developers are featured on trade press websites, are representative of the underlying issue.

7. While this text does not examine game developers from Blizzard (creator of *WoW*), there are likely more parallels between this text and Blizzard's working conditions than those of Linden Labs (which created *Second Life*).

8. While I draw out conceptually what I mean by the state in World 7, the foundation is located in a Deleuzian notion of the state apparatus based on the "abstract machine of overcoding" that it "tends increasingly to identify with the abstract machine it effectuates" (Deleuze and Guattari 1987, 223). So one can, for the time being, substitute "the state apparatus" for "the state" in my writing. While I largely conceptualize of it in a Gramscian sense of "coercion" and "consent" (Simon 2001, 24–32), I am particularly interested in the moment where the "perogative" power of the state is mobilized. I am interested in the "'legitimate' arbitrary aspect" or "extralegal, adventurous, violent" aspects of the state (Brown 1995, 186); the moment when consent transitions to coercion.

9. Greg Costikyan is an interesting figure in the videogame industry, who began as a board game developer and has worked on numerous videogame projects. He has also been a caustic and amusing commentator on the videogame industry. See his website at http://www.costik.com.

10. World 5 discusses further how passion lends the rant some of its credibility.

11. The name itself, "Zero Punctuation," is indicative of the kind of delivery rants take. When in the heat of passion, the ranter can hardly muster the effort needed to breathe. If the ranter did not care, he or she could not garner the exhaustive energy needed for delivering an impassioned speech desperate for change to the videogame industry.

12. The first essay I ever produced for publication while in graduate school, was an essay desperate for finding new kinds of rhetorics and metaphors by which to frame critical endeavors that did not appeal to the kind of militaristic analogies that dominate much of science and technology studies (O'Donnell 2004b).

13. The game industry had over $22 billion dollars in sales in 2008 and $19.6 billion in 2009 (Remo 2010; ESA 2009). These figures do not include numbers for videogame rentals, the sale of used videogames, or money made from the licensing of videogame intellectual properties (IP) to movie companies.

World 2

1. And so, too, is it important for researchers interested in studying the industry to be acquainted with these languages. Without this knowledge it would have been impossible to gain access to my initial field site. It was actually a conversation about game development and project planning at a party that began the friendship that provided me the opportunity to do pilot research at a game studio.

2. Even the few game developers who did not consider themselves gamers prior to working in the game industry have at least one of the latest console systems. They try to play, at least casually, the latest and most popular game titles.

3. On the Nintendo Entertainment System's controller, this sequence was up, up, down, down, left, right, left, right (on the directional pad) and the buttons "B" then "A" followed by pressing the "Start" button.

4. A game mechanic is the underlying "game" which is then presented on a screen. Think of it as the rule that higher cards beat lower cards, and equal cards mean war in the card game War.

5. *Spy vs. Spy* was originally a *Mad Magazine* comic strip, which was later inspiration for a videogame on several different console game platforms. It is a game with a long history of its own. The mechanic, which interested my informants, was the idea that

rather than direct combat (i.e., shooting or punching your opponent), users could set traps for one another.

6. It's possible that this demand for at least a kind of "gamer" identity limits the accessibility of game design as a profession. Certainly, no one would question that knowledge of games is not crucial for game designers. Yet, the identification of one's self as "gamer," may limit the number of people interested in pursuing it as a career. Certainly there is no similar demand put on those interested in making films. They may care for and be very knowledgeable of the world of filmmaking. Yet, they would not likely call themselves "filmies" or "filmers."

7. The idea that "design is thus best seen as a process of communication, negotiation, and consensus-building," (Bucciarelli and Kuhn 1997, 214) has been useful in my research, especially in the context of the videogame industry because it involves so much communication and negotiation. The idea of design as a consensus-building project also dramatically complicates the idea that the designers are solely responsible for the final game, which may have been shaped by dozens of other forces.

8. Time spent in front of the screen can be spent working on creating models based off of drawings, taking existing models and modifying them for new purposes, changing textures or the ways in which a texture is distributed over a model, or the way a model has been animated. Computer time is also spent dealing with IMs, email, web browsing, all work (related and unrelated).

9. "Bones" are the underlying elements that allow 3D models to be animated. Similar, conceptually, to the bones of a person, these objects are used to animate a 3D model and control how it is capable of being moved. However, the "mesh" of the model, whose motion is dictated by the bones, is not governed by some mysterious method, artists must ensure that the motion looks adequate and may have to use fewer virtual bones than, perhaps, the "real thing" would contain.

World 3

1. "Golden master" refers to the state of a game when it is ready to be shipped to the manufacturing company for mass production.

2. In part I see the lack of change in how disciplines work together as connected with the continued separation of disciplines among game developers. It matters which group you are working with. Are you an engineer, designer, artist, or manager? The setting off of groups from other groups creates a "sort of Mafia," or "inbred group of buddies," who "do things the way the want" (Fortun 2001, 116). This exacerbates the differences and separations between groups, which matters when they are constantly interacting and working within and among the elements that another group creates.

3. "Collision data" being stored separately from the model geometry of a level is one of the many contextual game development practices that are never shared more broadly. Storing collision data separately allows for faster or "cheaper" computation of collision detection. Rather than using the level geometry, simpler objects, such as a sphere or box can represent more complex objects.

4. I must admit my predilection toward the tools engineer, having been one when I was a game developer. I will attempt to moderate my awe of this position when possible. I suppose it only makes sense, given that my transition to this role went along with a decision to pursue graduate studies that led to this work.

5. As long as I'm making confessions, I should probably admit my residual frustration with console manufacturers as well. This frustration was really driven home as I worked with my Chennai, India-based informants on a game that was later canceled and the studio closed. I wondered, why they were making similar missteps that I had made six years earlier in my work in the industry? What kept me from sharing that information? The answer is quite simple there: NDAs and proprietary hardware.

6. The term "management" is used to reference these as a collective game development endeavor.

7. This seems to contradict the idea that work though, "increasingly individualized… is disaggregated in its performance, and reintegrated in its outcome through a multiplicity of interconnected tasks in different sites, ushering in a new division of labor based on the attributes/capacities of each worker rather than on the organization of the task" (Castells 1998, 502). Furthermore, based on my experiences, the unfortunate side effect of this has been increased time at work to make up for the re-socialized workplace. This further contradicts the idea that "skilled labor is required to manage its own time in a flexible manner, sometimes adding more work time, at other times adjusting to flexible schedules, in some instances reducing working hours, and thus pay" (Castells 1998, 468). Such an arrangement would seem an outrageous luxury in game development.

World 4

1. US-based software companies have assumed that emerging economies would automatically adopt the same tools as their predecessors, no matter the cost. This complicates the matter significantly for Indian studios, who frequently cannot afford the combined price tag of Max, Visual Studio, various Adobe products, and Perforce. This frequently leaves them using less expensive, though more difficult methods such as shared folders or SVN. The need for simplified tools has also provided entrepreneurial opportunities for Indian development firms, for some studios have developed their own systems for solutions with an eye toward commoditizing them. The extremely high cost of software combined with global differentials in

money markets has driven foreign companies to develop their own technologies, several of which may eventually compete with those of US companies.

2. This is in contrast with engineers who, when necessary, can work simultaneously and "merge" their efforts later.

3. The "mangle" of game development (Pickering 1995, 7) is in part a product of the dance with hardware systems that may or may not work as advertised. This is further complicated by "the world. . . . continually doing things" (Pickering 1995, 7–8) and these things, such as electrons moving through circuitry, media devices spinning up, power flowing out of a batter, are frequently mediated or "threaded" through technological devices (Pickering 1995, 7–8). The situation is even more complicated if your devices are highly unpredictable and not necessarily documented.

4. This "creative ambiguous process" or the necessity of "intellectual flexibility" is empowering for many, but also places the onus of production on the individual. If unable to produce, then they are assumed simply unskilled or just not smart enough. In many cases, there is a complex relationship between the context of work being done and those forces that enable or constrain it. While the "desired results or functions are what are demanded of workers, the contextual mechanisms, by which output is reached, while often dictated, frequently has little to do with the actual means by which things occur" (English-Lueck and Saveri 2001, 8). Workers are often judged based upon dictated demands rather than on the contextualized mechanisms necessary to actually do the work. Put another way, sometimes it really isn't your fault that things keep exploding, but rather the set of moving targets that encourages those catastrophes.

5. Though perhaps far removed from a chemical plant in Bhopal, the continued systematic failure of breakdown of game development practices seems to me to indicate a problem more systemic than user error (Fortun 2001, 123–131). The "modifications" and ad hoc modifications of complex technical systems can have unforeseen results, something game developers can certainly understand. Furthermore, as the complexity or coupling of a system increases, the opportunity for "catastrophe" or "system accident" increases rapidly (Perrow 1999, 62–100).

6. The connection to Development Kits or DevKits, explained in more detail in World 6, is an important one. These complex technological systems are supported by complex software systems and custom processes must couple together to ensure the overall health of the build.

7. What game developers need just as much as "interactive" systems is better processes for pursuing their assemblages. Much like other science industries, game development is "such a dense, intricate, and volatile assemblage of practices, metaphors, articulations, and other kludged-together elements of nature, culture, and power, they have to be muddled through." But more important, this process must

remain "cautious, nimble, and respectful, since they deal with explosive matter" (Fortun and Bernstein 1998, 147–148).

World 5

1. In her research on open source software developers, Coleman describes the deep hack mode as "a cavernous state of mental and often physical isolation in which one reaches such a pure state of concentration that basic biological drives like sleeping and eating are put on hold during the hours or days that pass" (Coleman 2001, 233). Deep hack mode is much like a kind of "flow" state (Csikszentmihalyi 2008) that many game players find themselves in when immersed in a play experience.

2. The full video can be viewed at https://www.youtube.com/watch?v=LkCNJRfSZBU.

3. The "32.33 percent repeating" chance of survival actually comes from the dialogue of the Jenkins video clip. Because the clip was actually staged and meant to be humorous, I can only surmise that the "0.33 percent repeating" component was meant as a joke to poke fun at *WoW* players or gamer "nerds" more generally.

4. Game developers certainly "muddle through" the "socially complicated as well as intellectually complex" process of creating their technological systems (Fortun and Bernstein 1998, x–xi). What I think differentiates the game industry is that it as of yet has no systematic system for reporting, publicizing, or thinking more broadly about that process of muddling. While tension remains between talking about and documenting the process of scientific production, the difference in the game industry is that there is no broader discussion, not even an opportunity for tension.

5. Session listings were previously available, online, though can now be searched as part of the GDC "vault" project (www.gdcvault.com). In 2007, five talks with agile in their title were presented, six in 2008, one in 2009, and 11 in 2010 (GDC 2007; GDC 2008; GDC 2009; GDC 2010). More contained the word "process" or "management" with often the focus of these talks examining art asset production pipelines and talent pool management.

6. Latour writes about "technical skill" as applying to those "with a unique ability, a knack, a gift, and also to the ability to make themselves indispensable, to occupy privileged though inferior positions, which might be called . . . obligatory passage points" (Latour 1999, 191). Though I do not think this concept of giftedness is entirely applicable to game development, it encourages critical thinking about the work, in which concepts like "passion," "skill," and "talent," seem to imply innate capacities.

7. An infamous example can be found at http://www.collegehumor.com/video/949720/tighten-up-the-graphics.

8. This is in direct contradiction of what many refer to as a "mature" industry. While I would agree with an assessment of the industry as "structured" (Williams

2002, 51), maturity or stability has not been reached in the game development industry. What I think can be mistaken for "maturity" is rather what the same author identifies as the "entry barrier created by an existing dominant network" (Williams 2002, 51).

9. Nearly a year and a half after the publication of the ea_spouse blog, Erin Hoffman, now a designer in the videogame industry, made her identity public in an online interview (Wong 2006). During the course of my research, she began working with a game studio that spun off from my primary field site. She continues to write for online game development and game-related publications and more recently has released her first novel as a fantasy writer.

World 6

1. Several texts have mentioned the existence of the 10NES chip, but none of these texts ever offered any proof of the existence of the device (Clapes 1993; Sheff 1993). Despite this, others have been willing to use these reports as facts, without inquiry into the validity of these claims or the functionality of the device (Kline et al. 2005).

2. While I think the situation is more complicated, as this world and much of the text would seem to indicate, I do believe that viewing code/technology as "legislation" as well as "speech" is a productive tool for thinking about technology (Lessig 1999). Taken to its extreme however, I think it assumes too much about the deterministic character of technology.

3. There are, of course, examples of companies that managed to circumvent Nintendo's lockout mechanisms. As near as I can tell based on unscientific searches, 87 unlicensed titles have been released, compared with 670 licensed titles (Nintendo 2003).

4. As the following court case selection indicates, Atari's resulting silicon lock pick was called the "rabbit program." Though I use the term "harebrained," I do so with tongue in cheek, the kind of ironic wordplay that would not be uncommon in game development companies. I do not intend it simply to mean "stupid."

5. I actually construct this in contradiction to accounts of neoliberalism as a hollowing out of the state, or the "subjugation of political and social life to a set of processes termed 'market forces'" (Farmer 2005, 5) or the advocacy of "a competition-driven market model" (Farmer 2005). Rather, the codifying of corporate power seems a mobilization of the state to perform duties that corporations operating in a market cannot do. In many respects it is making the move for both de-regulation and increased regulation of political and social life.

6. Or in a Deleuzian way, "This whole chain and web of power is immersed in a world of mutant flows that eludes them. It is precisely its impotence that makes power so dangerous. The man of power will always want to stop the lines of flight,

and to this end to trap and stabilize the mutation machine in the overcoding machine" (Deleuze and Guattari 1987, 229). This is precisely why control must be maintained over the realm of the producer. Left to its own devices it will move in directions that may make SEC filings more difficult, but perhaps more beneficial to new markets and new producers and competitors.

7. The "DevKit" is distinct from "development kits" as defined by some authors (Postigo 2003, 603). There is a slippery and important language to keep in mind. SDKs or software development kits are distinctly different, though intertwined with DevKits. DevKits typically have accompanying SDKs. However, it is possible for companies to release SDKs without having DevKits. The hardware of the DevKit is in part what distinguishes it from an SDK, as does access to documentation and other resources like online discussion forums.

8. Also strikingly uncanny are current "trusted computing" endeavors by PC and software companies that I see making an appeal to the 10NES's pattern of game development where you can be much more sure that a user has paid for what they are playing: game, song or otherwise. "The 'trusted' part of this system is that this device obeys rules established by the copyright owner when they first make the song available. . . . The rhetoric is classic command-and-control, a far cry from the delicate balance of copyright" (Gillespie 2004, 241).

9. Because of the peculiarities of videogame consoles, game developers must continually re-create common software that, even if ported to a console cannot be released back to the community.

World 7

1. A console manufacturer will not distribute an unrated game, even if the producer has managed to bypass the production control mechanisms implemented in the console. For such a product, if distributed, subjects creators to criminal consequences under the jurisdiction of the Digital Millennium Copyright Act (DMCA).

2. Since the controversy surrounding the sharing of music files on Napster, music files have dominated the public awareness of where DRM and the DMCA impact their lives (Gillespie 2007). Unfortunately the impact is far broader, and much of the emphasis remains on the technology creators rather than the copyright holders who are most often responsible for making the digital lock-down demands in the first place.

3. While I treat the distinction between hackers and crackers as relatively clear, it is actually a rather complex division, even among hackers (Coleman 2005, 53–54). In part I do so as a reaction to the alternatively muddy conflation, which is frequently tossed about in which all hackers are bad. This erring on the side of upstanding hacking activity is done with the reasonable assumption that not every hacker has criminal intent.

4. While it is still possible that some day in the future this project will be dismantled by Nintendo, at this point it seems unlikely. The project can currently be found at http://sourceforge.net/projects/vba, with tools available for most PC platforms, including Windows, Mac OS X, and Linux.

World 8

1. Oddly, a game industry game already exists. *Game Dev Story* was released for Apple's iOS platform in October of 2010. This resource management and strategy game challenges players to manage a game development studio. The game takes small jabs at numerous aspects of the game industry, from console manufacturers to publishers and game titles. Of particular note is that fact that the focus remains at the studio level even as developers level up and improve as they remain with your company. The game developers themselves disappear into the background, functioning as yet another resource to be managed. And the game reinforces this perspective; the procedural rhetoric of the game marginalizes the labor of game developers, yet outside the console, game developers are the culture and creative collaborative community that makes the game industry function in *Game Dev Story*, rather than forgettable resources. In contrast, the game presented here in World 8 focuses on the individuals working within the industry.

2. Experienced game developers will, of course, now shudder, "That's not a design document! You haven't even figured out what all those puzzles are going to be!" Yes, you are right. The engineers will be shaking their heads and mumbling something about, "How are you going to specify those puzzles and their win/fail states?" Artists will be worried about the lack of concept art and a real visual direction. But those developers also know that a complete design for such a game would be a text significantly longer than appropriate for a single chapter of this book. I take your critique seriously, however.

3. I have written at length about the role that game development tools have in the design and creation of videogames. This is done through the lens of a series of patents by Nintendo that outlined the technological foundations for a truly co-creative production platform for games, affectionately dubbed *Mario Factory* (O'Donnell 2013).

Glossary

2D (Two-dimensional). Flat graphics meant to be viewed like an image or flat text.

3D (Three-dimensional). Graphics meant to be viewed within a three-dimensional space.

AAA (Triple A). Drawn from the sporting world, where AAA was the designation of the "Big Leagues." Imported into the game industry to designate those working on "big" games.

AI (Artificial intelligence). Logic provided to interactive elements within a game not under human control.

Agile. A software development methodology/ideology.

Assets. The art (models, textures, audio, etc.) information that a game engine "reads" in, in order to present a game to the user.

Console (Debug). A text-based command system for adjusting the underlying state of a game's engine. The debug console is most frequently removed or hidden when a game is shipped.

Console (Game). A self-contained computing system typically connected to a television. Game consoles frequently have custom hardware, software, and input devices to differentiate themselves from their competitors.

Data. The "design" data, which can range from scripts to structured information like XML, which is used by the engine to further define the game.

DevKits (Development Kits). The hardware and software of a game console development kit. Necessary to develop a game for a given piece of console hardware.

DRM (Digital Rights Management). Copy protection mechanism.

DS (Nintendo DS). A handheld game console manufactured by Nintendo.

Engine. The underlying software of a game. The engine is not a game; it is more basic than that. It provides a platform, to which more code, data, and art assets are produced to make a game.

ESA (Entertainment Software Association). Body that lobbies on the part of large game studios, publishers, and manufacturers.

ESRB (Entertainment Software Rating Board). Body that rates, for a fee, games for retail sale in the United States.

F/LOSS (Free/Libre Open Source Software). Software which can be freely licensed for use and which the underlying source code is available to the licensee.

GameCube. A game console manufactured by Nintendo.

GDC (Game Developers Conference). The annual conference, typically held in San Francisco, to which game developers who are able to flock to each year.

HUD (Heads-up display). The user-interfaces typically always visible in a game. These often provide information regarding the current state of a game, ranging from location to health or points.

ICE (Immigration and Customs Enforcement). U.S. Agency that has been involved with raids related to MOD chips.

IDE (Integrated Development Environment). The software tools used by programmers to edit, debug, and compile code.

IGDA (International Game Developers Association). A volunteer-based professional organization that works to represent the interests of game developers more broadly.

ISP (Internet service provider). Comcast, Verizon, and AT&T are examples of ISPs.

Konami Code. A special sequence of moves on a game-pad controller that provided the player with extra lives in the game *Contra* on the NES.

Max (3D Studio Max). A professional 3D modeling, rigging, and animation application developed by Autodesk.

Maya (Maya). A professional 3D modeling, rigging, and animation application developed by Autodesk.

Middleware. Software that does not itself constitute a game, but that can be used to ease the development time or provide functionality to a game. A game engine, a physics engine, and a sound engine are examples.

MOD (Modification). A stand-alone modification to a game that can be shared or redistributed that changes the underlying functionality or behavior of a game engine.

NDA (Non-disclosure agreement). A legal document requiring one or more parties to not disclose secret information to others. What precisely is secret and who others is up for grabs.

NES (Nintendo Entertainment System). A game console released by Nintendo in the United States in 1984.

Polygons. The triangles or "faces" that make up a 3D model.

Preproduction. The early phases of game development, spent trying to "find" the game.

Production. The phase of game development where art and design data production is dramatically increased.

Programming languages. These can be either "compiled" languages like C or C++ as well as "interpreted" scripting languages like Lua, Python, or others.

PSP (Playstation Portable). A game console made by Sony Computer Entertainment.

Q/A (Quality assurance). The testing and usability component of game development.

QoL (Quality of life). The "well-being" of people in a given context.

Scrum. An element of or style of Agile development.

SDK (Software development kit). The software associated with either a development environment or platform.

SM3 (Spiderman 3). A movie and a videogame for a variety of game consoles. The primary project observed throughout this text.

SVN (Subversion). A VCS.

Textures. The visual skins found on 3D models.

VCS (Version control system). Not to be confused with the Atari VCS, an early game console. The VCS in this text is a means for sharing and "versioning" or monitoring the revisions to a file.

VS (Vertical slice). The prototype phase of game development where a "slice," as in a slice of cake, of a game's visuals and gameplay are produced in order to demonstrate playability and feasibility.

VV (Vicarious Visions). The game development studio that for one reason or another was willing to let this anthropologist spend nearly three and a half years within its halls.

Wii (Nintendo Wii) A game console made by Nintendo.

WoW (World of Warcraft) A massively multiplayer online (MMO) world developed by Blizzard.

Xbox (Microsoft Xbox). A game console made by Microsoft.

Xbox 360 or 360 (Microsoft Xbox 360). A game console made by Microsoft.

XML (Extensible Markup Language). A structured markup language, like HTML, which is frequently used to organize data within games or other software packages.

References

Adorno, Theodor W., and Max Horkheimer. 1976. *Dialectic of Enlightenment*. Translated by John Cumming. New York: Continuum International Publishing Group.

Alexander, Leigh. 2010. "Analysis: Is the Game Industry a Happy Place?" Gamasutra. Accessed July 28, 2010. http://www.gamasutra.com/view/news/29292/Analysis_Is_The_Game_Industry_A_Happy_Place.php.

Androvich, Mark. 2007. "Sony Threatens to Pursue Legal Action against PS3 Hackers." GamesIndustry International. Accessed July 20, 2007. http://web.archive.org/web/20070706161709/http://www.gamesindustry.biz/content_page.php?aid=25750.

Appadurai, Arjun. 1996. *Modernity at Large*. Minneapolis: University of Minnesota Press.

Atari, Tengen, and Nintendo. 1992. *Atari Games Corp. and Tengen, Inc. v. Nintendo of America Inc. and Nintendo Co., Ltd.*, 975 F.2d, 832.

Baba, Marietta L. 2003. "Working Knowledge Goes Global: Knowledge Sharing and Performance in a Globally Distributed Team." *Anthropology of Work Review* 24 (1–2): 19–29.

Barley, Stephen R. 1996. "Technicians in the Workplace: Ethnographic Evidence for Bringing Work into Organizational Studies." *Administrative Science Quarterly* 41 (3): 404–441.

Barley, Stephen R., and Julian E. Orr. 1997a. *Between Craft and Science: Technical Work in U.S. Settings*. Ithaca, NY: Cornell University Press.

Barley, Stephen R., and Julian E. Orr. 1997b. "Introduction: The Neglected Workforce." In *Between Craft and Science: Technical Work in U.S. Settings*, ed. Stephen R. Barley and Julian E. Orr, 1–19. Ithaca, NY: Cornell University Press.

Bates, Bob, Jason Della Rocca, Alex Dunne, John Feil, Mitzi McGilvray, Brian Reynolds, Jesse Schell, and Kathy Schoback. 2004. "Quality of Life Issues." IGDA. Accessed

May 20, 2014. https://web.archive.org/web/20041119062234/http://www.igda.org/qol/open_letter.php.

Beck, Ulrich. 2000. *The Brave New World of Work*. Malden, MA: Polity Press.

Becker, Howard. 1984. *Art Worlds*. Berkeley: University of California Press.

Bogost, Ian. 2006. "Persuasive Games: Wii's Revolution Is in the Past." CMP Serious Games Source. Accessed July 20, 2007. http://seriousgamessource.com/features/feature_112806_wii_1.php.

Bogost, Ian. 2007. *Persuasive Games: The Expressive Power of Videogames*. Cambridge, MA: MIT Press.

Bogost, Ian. 2011. "'Gamification Is Bullshit.'" The Atlantic. Accessed August 29, 2011. http://www.theatlantic.com/technology/archive/2011/08/gamification-is-bullshit/243338.

Bonds, Scott, Jamie Briant, Dustin Clingman, Hank Howie, François Dominic Laramée, Greg LoPiccolo, Andy Luckey, and Mike McShaffry. 2004. *Quality of Life in the Game Industry: Challenges and Best Practices*. Mount Royal, NJ: International Game Developers Association.

Bowker, Geoffrey C., and Susan Leigh Star. 1999. *Sorting Things Out: Classification and Its Consequences*. Cambridge, MA: MIT Press.

Brooks, F. P. 1995. *The Mythical Man-Month: Essays on Software Engineering*. Boston, MA: Addison-Wesley.

Brown, Wendy. 1995. *States of Injury: Power and Freedom in Late Modernity*. Princeton, NJ: Princeton University Press.

Bucciarelli, Louis L. 1994. *Designing Engineers*. Cambridge, MA: MIT Press.

Bucciarelli, Louis L., and Sarah Kuhn. 1997. "Engineering Education and Engineering Practice: Improving the Fit." In *Between Craft and Science: Technical Work in U.S. Settings*, edited by Stephen R. Barley and Julian E. Orr, 210–229. Ithaca, NY: Cornell University Press.

Burghardt, Gordon M. 2005. *The Genesis of Animal Play: Testing the Limits*. Cambridge, MA: MIT Press.

Callon, Michel. 1989. "Society in the Making: The Study of Technology as a Tool for Sociological Analysis." In *The Social Construction of Technological Systems: New Directions in the Sociology and History of Technology*, edited by Wiebe Bijker, Thomas P. Hughes, and Trevor Pinch, 83–106. Cambridge, MA: MIT Press.

Carless, Simon. 2007. "Why Consoles Are Here to Stay, Yay." GameSetWatch. Accessed August 4, 2007. http://www.gamesetwatch.com/2007/08/why_consoles_are_here_to_stay.php.

Carter, Ben. 2003. "Postmortem: Lost Toys' Battle Engine Aquila." *Game Developer Magazine* 10 (4): 50–58.

Cassell, Justine, and Henry Jenkins. 2000. *From Barbie to Mortal Kombat: Gender and Computer Games*. Cambridge, MA: MIT Press.

Castells, Manuel. 1998. *The Rise of the Network Society*. Oxford, UK: Blackwell Publishers.

Castronova, Edward. 2005. *Synthetic Worlds: The Business and Culture of Online Games*. Chicago: University of Chicago Press.

Chaplin, Heather, and Aaron Ruby. 2005. *Smartbomb: The Quest for Art, Entertainment, and Big Bucks in the Videogame Revolution*. Chapel Hill, NC: Algonquin Books.

Chen, Brian X. 2008. "iPhone Developers Devote Profane Web Site to Apple's NDA." *Wired*. Accessed August 6, 2008. http://www.wired.com/gadgetlab/2008/08/developers-stea.

Clapes, Anthony L. 1993. *Softwars: The Legal Battles for Control of the Global Software Industry*. Westport, CT: Quorum Books.

Clifford, James. 1986. "On Ethnographic Allegory." In *Writing Culture: The Poetics and Politics of Ethnography*, edited by James Clifford and George E. Marcus, 98–121. Berkeley: University of California Press.

Coleman, Gabriella E. 2001. "High-Tech Guilds in the Era of Global Capital." *Anthropology of Work Review* 22 (1): 28–32.

Coleman, Gabriella E. 2005. "The Social Construction of Freedom in Free and Open Source Software: Hackers, Ethics, and the Liberal Tradition." Diss., Department of Anthropology, University of Chicago, Chicago, IL.

Consalvo, Mia. 2007. *Cheating: Gaining Advantage in Videogames*. Cambridge, MA: MIT Press.

Coupland, Douglas. 2006. *JPod*. New York: Bloomsbury Publishing.

Csikszentmihalyi, Mihaly. 2008. *Flow: The Psychology of Optimal Experience*. New York: Harper and Row.

Danks, Mark. 2008. "PlayStation-edu." Playstation.blog. Accessed May 16, 2010. http://blog.us.playstation.com/2008/06/06/playstation-edu/.

Datamonitor. 2005. "Games Consoles in the United States."

Datamonitor. 2007. "Games Consoles Industry Profile: United States."

Datamonitor. 2004. "Games Consoles in the United States."

Davis, Galen. 2006. "GDC Rant Heard 'Round the World." Gamespot News. Accessed September 14, 2006. http://www.gamespot.com/news/6120449.html.

Deleuze, Gilles, and Felix Guattari. 1987. *A Thousand Plateaus: Capitalism and Schizophrenia*. Translated by Brian Massumi. Minneapolis: University of Minnesota Press.

Deuze, Mark. 2007. *Media Work*. Malden, MA: Polity Press.

Deuze, Mark, Chase Bowen Martin, and Christian Allen. 2007. "The Professional Identity of Gameworkers." *Convergence* 13 (4): 335–353.

DMCA. 1998. To amend title 17, United States Code, to implement the World Intellectual Property Organization Copyright Treaty and Performances and Phonograms Treaty, and for other purposes, 17 U.S.C. §§ 512, 1201–1205, 1301–1332; 28 U.S.C. § 4001; 17 U.S.C. §§ 101, 104, 104A, 108, 112, 114, 117, 701.

Dobson, Jason. 2007. "Nintendo Issues 'Strong Support' for U.S. Anti-Piracy Measures." Gamasutra.com. Accessed July 20, 2007. http://www.gamasutra.com/php-bin/news_index.php?story=13466.

Downey, G. L. 1998. *The Machine in Me: An Anthropologist Sits among Computer Engineers*. New York: Routledge Press.

Downey, Greg. 2001. "Virtual Webs, Physical Technologies, and Hidden Workers." *Technology and Culture* 42 (2): 209–235.

Dyer-Witheford, Nick. 1999. "The Work in Digital Play: Video Gaming's Transnational and Gendered Division of Labor." *Journal of International Communication* 6 (1): 69–93.

Dyer-Witheford, Nick, and Greig de Peuter. 2006. "EA Spouse and the Crisis of Video Game Labour: Enjoyment, Exclusion, Exploitation, and Exodus." *Canadian Journal of Communication* 31 (3): 599–617.

Dyer-Witheford, Nick, and Greig de Peuter. 2009. *Games of Empire: Global Capitalism and Video Games*. Edited by Katherine Hayles, Mark Poster, and Samuel Weber. Minneapolis: University of Minnesota Press.

Dyer-Witheford, Nick, and Zena Sharman. 2005. "The Political Economy of Canada's Video and Computer Game Industry." *Canadian Journal of Communication* 30 (2).

ea_spouse. 2004. "EA: The Human Story." *Live Journal*. Accessed June 24, 2010. http://ea-spouse.livejournal.com/274.html.

Edery, David J. 2007. "Console Demise? Don't Hold Your Breath." Game Tycoon. Accessed July 23, 2007. http://www.edery.org/2007/07/console-demise-dont-hold-your-breath.

Electronic Arts, Inc. 2007. "Form 10-Q: Quarterly Report Pursuant to Section 13 or 15 (d) of the Securities Exchange Act of 1934."

English-Lueck, J. A. and Andrea Saveri. 2001. "Silicon Missionaries and Identity Evangelists." *Anthropology of Work Review* 22 (1): 7–12.

ESA (Entertainment Software Association). 2005. "2005 Sales, Demographics, and Usage Data: Essential Facts about the Computer and Video Game Industry." Entertainment Software Association. Accessed February 5, 2008. http://web.archive.org/web/20080121185227/http://www.theesa.com/files/2005EssentialFacts.pdf.

ESA (Entertainment Software Association). 2009. "2009 Sales, Demographics, and Usage Data: Essential Facts about the Computer and Video Game Industry." Entertainment Software Association. Accessed May 20, 2014. http://www.theesa.com/facts/pdfs/esa_ef_2009.pdf.

Evans, Geoff. 2008. "Nocturnal Initiative." Insomniac Games. Accessed May 25, 2010. http://nocturnal.insomniacgames.com.

Fahey, Rob. 2007. "Locked Away: Do the Death Throes of Music DRM Mean Anything for Games?" GamesIndustry.Biz. Accessed May 1, 2007. http://web.archive.org/web/20070616130240/http://www.gamesindustry.biz/content_page.php?aid=24585.

Farmer, Paul. 2005. *Pathologies of Power: Health, Human Rights, and the New War on the Poor*. Edited by Robert Borofsky. Berkeley: University of California Press.

Feil, John, and David Weinstein. 2006. "Game Industry Crediting: A Snapshot of the Present." IGDA. Accessed March 22, 2007. http://www.igda.org/credit/IGDA_CreditsSnapshot_Apr06.pdf.

Finley, Alyssa. 2007. Postmortem: 2K Games' Bioshock. *Game Developer Magazine* 14 (10): 20–26.

Forsythe, Diana E. 2001. *Studying Those Who Study Us: An Anthropologist in the World of Artificial Intelligence*. Edited by Timothy Lenoir and Hans Ulrich Gumbrecht. Stanford, CA: Stanford University Press.

Fortun, Kim. 2001. *Advocacy after Bhopal: Environmentalism, Disaster, New Global Orders*. Chicago: University of Chicago Press.

Fortun, Kim. 2006. "Poststructuralism, Technoscience, and the Promise of Public Anthropology." *India Review* 5 (3–4): 294–317.

Fortun, Mike, and Herbert J. Bernstein. 1998. *Muddling Through: Pursuing Science and Truths in the 21st Century*. Washington, DC: Counterpoint Press.

Friedman, Thomas L. 2005. *The World Is Flat: A Brief History of the Twenty-First Century*. New York: Farrar, Straus and Giroux.

Galison, Peter. 1997. *Image and Logic: A Material Culture of Microphysics*. Chicago: University of Chicago Press.

GDC. 2007. "Session Listings for GDC 2007." Accessed November 14, 2007. http://gdcvault.com/browse/gdc-07.

GDC. 2008. "Session Listings for GDC 2008." Accessed August 11, 2010. http://gdcvault.com/browse/gdc-08.

GDC. 2009. "Session Listings for GDC 2009." Accessed August 11, 2010. http://gdcvault.com/browse/gdc-09.

GDC. 2010. "Session Listings for GDC 2010." Accessed August 11, 2010. http://gdcvault.com/browse/gdc-10.

Gill, Rosalind. 2007. *Technobohemians or the New Cybertariat? New Media Work in Amsterdam a Decade after the Web*. Amsterdam: Institute of Network Cultures.

Gillespie, Tarleton. 2004. "Copyright and Commerce: The DMCA, Trusted Systems, and the Stabilization of Distribution." *Information Society* 20 (4): 239–254.

Gillespie, Tarleton. 2006. "Designed to 'Effectively Frustrate': Copyright, Technology, and the Agency of Users." *New Media & Society* 8 (4): 651–669.

Gillespie, Tarleton. 2007. *Wired Shut: Copyright and the Shape of Digital Culture*. Cambridge, MA: MIT Press.

Gourdin, Adam. 2005. "Game Developer Demographics: An Exploration of Workforce Diversity." International Game Developers Association. Accessed August 14, 2011. http://www.igda.org/game-developer-demographics-report.

Gramsci, A. 1975. *Prison Notebooks*. New York: Columbia University Press.

Hakken, David. 2000a. "Resocializing Work? Anticipatory Anthropology of the Labor Process." *Futures* 32: 767–775.

Hakken, David. 2000b. "Resocializing Work? The Future of the Labor Process." *Anthropology of Work Review* 21 (1): 8–10.

Hall, Stuart. 1996. "Gramci's Relevance for the Study of Race and Ethnicity." In *Critical Dialogues in Cultural Studies*, edited by David Morely and Kuan-Hsing Chen, 411–441. New York: Routledge.

Haraway, Donna Jeanne. 1997. *Modest_Witness-AT-Second_Millennium.FemaleMan©_Meets_OncoMouse™: Feminism and Technoscience*. New York: Routledge Press.

Harvey, David. 1990. *Condition of Postmodernity: An Enquiry into the Origins of Cultural Change*. Cambridge, MA: Blackwell Publishing.

Hoffman, Erin. 2009. "Why Your Game Idea Sucks." The Escapist Magazine. Accessed November 16, 2009. http://www.escapistmagazine.com/articles/view/issues/issue_221/6582-Why-Your-Game-Idea-Sucks.

Hughes, Thomas P. 1999. "Edison and Electric Light." In *The Social Shaping of Technology*, edited by Donald MacKenzie and Judy Wajcman, 50–63. Philadelphia, PA: Open University Press.

Huizinga, Johan. 1971. *Homo Ludens: A Study of the Play-Element in Culture*. Boston: Beacon Press.

Hyman, Paul. 2007. "For Better or Worse: A Quality of Life Update." *GameDeveloper Magazine*, June/July: 7–11.

ICE (U.S. Immigration and Customs Enforcement). 2007. "Game Over: ICE, Industry Team Up in Gaming Piracy Crackdown: 32 Search Warrants Executed in Nationwide Intellectual Property Rights Investigation." Accessed June 2, 2010. https://web.archive.org/web/20100528032701/http://www.ice.gov/pi/news/newsreleases/articles/070801washington.htm.

Informant. 2009a. "Forum Post: 'Re: What's the Point (of Homebrew)?'"

Informant. 2009b. "Forum Post: 'What's the Point (of Homebrew)?'"

Informant and Casey O'Donnell. 2007a. "Interview with Artist ART_DS_Ogre_1."

Informant and Casey O'Donnell. 2007b. "Interview with Designer DESIGN_LEAD_1."

Informant and Casey O'Donnell. 2006a. "Email—Re: Press Start."

Informant and Casey O'Donnell. 2006b. "Interview with Engineer ENG_Asylum."

Informant and Casey O'Donnell. 2005a. "Interview with Engineer ENG_DS_Spidey_1."

Informant and Casey O'Donnell. 2005b. "Interview with Engineering Group Manager ENG_GRP_MGR_1."

Informant and Casey O'Donnell. 2005c. "Field Notes amongst Engineers."

Informant and Casey O'Donnell. 2005d. "GDC 2006 Abstract Submission: Where Is My Make Art Button? Improving the Artist/Programmer Relationship."

Informant and Casey O'Donnell. 2005e. "Interview with an Engineering Lead."

James, E. C. 2012. "Witchcraft, Bureaucraft, and the Social Life of (US)Aid in Haiti." *Cultural Anthropology* 27 (1): 50–75.

Johns, Jennifer. 2006. "Video Game Production Networks: Value Capture, Power Relations, and Embeddedness." *Journal of Economic Geography* 6:151–180.

Jones, S. E., and G. K. Thiruvathukal. 2012. *Codename Revolution: The Nintendo Wii Platform*. Cambridge, MA: MIT Press.

Kazemi, Darius. 2007. "Breaking In: Then and Now." Blogger.com. Accessed August 7, 2007. http://tinysubversions.blogspot.com/2007/08/breaking-in-then-and-now.html.

Kazemi, Darius. 2011. "Exploitify." TinySubversions.com. Accessed September 1, 2011. http://tinysubversions.com/2011/05/exploitify.

Kelly, John D. 2006. *The American Game: Capitalism, Decolonization, World Domination, and Baseball*. Chicago: Prickly Paradigm Press.

Kent, Steven L. 2001. *The Ultimate History of Video Games: The Story Behind the Craze that Touched our Lives and Changed the World*. New York: Three Rivers Press.

Kline, Stephen, Nick Dyer-Witherford, and Greig de Peuter. 2005. *Digital Play: The Interaction of Technology, Culture, and Marketing*. Québec, Canada: McGill-Queen's University Press.

Knorr-Cetina, Karin D. 1983. "The Ethnographic Study of Scientific Work: Towards a Constructivist Interpretation of Science." In *Science Observed: Perspectives on the Social Study of Science*, edited by Karin D. Knorr-Cetina and Michael Mulkay, 115–140. Beverly Hills, CA: Sage Publications.

Knorr-Cetina, Karin D. 1999. *Epistemic Cultures: How the Sciences Make Knowledge*. Cambridge, MA: Harvard University Press.

Krahulik, Mike, and Jerry Holkins. 2007. "Old School." Penny Arcade. Accessed April 8, 2008. http://www.penny-arcade.com/comic/2007/12/03.

Lashinsky, Adam. 2007. Search and Enjoy (Cover Story). *Fortune Magazine* 155 (1): 70–82.

Latour, Bruno. 1987. *Science in Action: How to Follow Scientists and Engineers Through Society*. Cambridge, MA: Harvard University Press.

Latour, Bruno. 1991. "Technology Is Society Made Durable." In *A Sociology of Monsters: Essays on Power, Technology and Domination*, edited by John Law, 103–131. New York: Routledge Press.

Latour, Bruno. 1999. *Pandora's Hope: Essays on the Reality of Science Studies*. Cambridge, MA: Harvard University Press.

Latour, Bruno. 2004. "Why Has Critique Run out of Steam? From Matters of Fact to Matters of Concern." *Critical Inquiry* 30 (2): 225–248.

Latour, Bruno, and Steve Woolgar. 1986. *Laboratory Life: The Construction of Scientific Facts*. Princeton: Princeton University Press.

Law, John. 1989. "Technology and Heterogeneous Engineering: The Case of Portuguese Expansion." In *The Social Construction of Technological Systems: New Directions in the Sociology and History of Technology*, edited by Wiebe Bijker, Thomas P. Hughes, and Trevor Pinch, 111–134. Cambridge, MA: MIT Press.

Leigh Star, Susan, and James R. Griesemer. 1989. "Institutional Ecology, 'Translations,' and Boundary Objects: Amateurs and Professionals in Berkeley's Museum of Vertebrate Zoology, 1907–39." *Social Studies of Science* 19 (3): 387–420.

Lessig, Lawrence. 2005. *Free Culture: The Nature and Future of Creativity*. New York: Penguin Press.

Lessig, Lawrence. 1999. *Code and Other Laws of Cyberspace*. Jackson, TN: Basic Books.

Lévy-Strauss, Claude. 1962. *The Savage Mind*. Edited by Julian Pitt-Rivers and Ernest Gellner. Chicago: University of Chicago Press.

Lik-Sang.com. 2006. "Lik-Sang.com Out of Business due to Multiple Sony Lawsuits." Wayback Machine. Accessed May 20, 2014. http://web.archive.org/web/20070314000426/http://www.lik-sang.com.

Lyotard, Jean-François. 1984. *The Postmodern Condition: A Report on Knowledge*. Edited by Wlad Godzich and Jochen Schulte-Sasse. Translated by Geoff Bennington and Brian Massumi. Minneapolis: University of Minnesota Press.

Malaby, Thomas M. 2009. *Making Virtual Worlds: Linden Lab and Second Life*. Ithaca, NY: Cornell University Press.

Malliet, Steven, and Eric Zimmerman. 2005. "The History of the Video Game." In *Handbook of Computer Game Studies*, edited by Joost Raessens and Jeffrey Goldstein, 23–46. Cambridge, MA: MIT Press.

Martin, Emily. 1997. "Managing Americans: Policy, Work, and the Self." In *Anthropology of Policy: Perspectives on Governance and Power*, edited by Cris Shore and Susan Wright, 239–260. New York: Routledge.

McAllister, Ken S. 2004. *Game Work: Language, Power, and Computer Game Culture*. Tuscaloosa: University of Alabama Press.

Montfort, N., and I. Bogost. 2009. *Racing the Beam: The Atari Video Computer System*. Cambridge, MA: MIT Press.

Murata, Taku. 2007. "Postmortem: Final Fantasy XII." *Game Developer Magazine* 14 (7): 22–27.

Nakagawa, Katsuya. 1985. U.S. Patent No. 4799635. Washington, DC: U.S. Patent and Trademark Office. Assignee: Nintendo Co., Ltd.

Nakagawa, Katsuya, and Masayuki Yukawa. 1987. U.S. Patent No. 4865321. Washington, DC: U.S. Patent and Trademark Office. Assignee: Nintendo Co., Ltd.

Nardi, Bonnie A. 2010. *My Life as a Night Elf Priest: An Anthropological Account of World of Warcraft*. Ann Arbor: University of Michigan Press.

Neff, Gina. 2012. *Venture Labor: Work and the Burden of Risk in Innovative Industries*. Cambridge, MA: MIT Press.

Neff, Gina. 2005. "The Changing Place of Cultural Production: The Location of Social Networks in a Digital Media Industry." *The Annals of the American Academy of Political and Social Science* 597: 134–152.

Neff, Gina, and David Stark. 2004. "Permanently Beta: Responsive Organization in the Internet Era." In *Society Online: The Internet in Context*, edited by Philip N. Howard and Steve Jones, 173–188. Thousand Oaks, CA: Sage Publications.

Neff, Gina, Elizabeth Wissinger, and Sharon Zukin. 2005. "Entrepreneurial Labor among Cultural Producers: 'Cool' Jobs in 'Hot' Industries." *Social Semiotics* 15 (3): 307–334.

Nintendo. 2003. "NES Licensed Game List." Nintendo. Accessed June 6, 2007. https://web.archive.org/web/20070317023021/http://www.nintendo.com/doc/nes_games.pdf.

Nintendo Co. LTD. and Nintendo of America Inc. v. Lik Sang International, LTD. 2003. HCA 3584/2002.

O'Donnell, Casey. 2004a. "A Case for Indian Insourcing: Open Source Interest in IT Job Expansion." *First Monday* 9.11. Accessed May 20, 2014. http://www.firstmonday.org/ojs/index.php/fm/article/view/1188.

O'Donnell, Casey. 2004b. "Critique with a "K"—From Military Matters of Concern to Cat's Cradles that Matter: In Search of Metaphors that Frame Inquiry." *Technoscience: Newsletter of the Society for Social Studies of Science* 19 (2): 1–3.

O'Donnell, Casey. 2008. "The Work/Play of the Interactive New Economy: Video Game Development in the United States and India." Ph.D. diss., Rensselaer Polytechnic University, Troy, NY.

O'Donnell, Casey. 2011a. The Nintendo Entertainment System and the 10NES Chip: Carving the Videogame Industry in Silicon. *Games and Culture* 6 (1): 83–100.

O'Donnell, Casey. 2011b. "Games Are Not Convergence: The Lost Promise of Digital Production and Convergence." *Convergence* 17 (3): 271–286.

O'Donnell, Casey. 2012. "This Is Not a Software Industry." In *The Video Game Industry: Formation, Present State and Future*, edited by P. Zackariasson and T. L. Wilson, 17–33. New York: Routledge.

O'Donnell, Casey. 2013. "Wither Mario Factory? The Role of Tools in Constructing (Co)Creative Possibilities on Videogame Consoles." *Games and Culture* 8 (3): 161–180.

Omi, Michael, and Howard Winant. 1994. *Racial Formation in the United States: From the 1960s to the 1990s*. New York: Routledge.

Orland, Kyle. 2008. "Sony Offers PS2/PSP Dev Kits for Education." Joystiq. Accessed May 16, 2010. http://www.joystiq.com/2008/06/06/sony-offers-ps2-psp-dev-kits-for-education.

Orr, Julian E. 1991. "Contested Knowledge: Work, Practice and Technology." *Anthropology of Work Review* 12 (3): 12–17.

Orr, Julian E. 1996. *Talking about Machines: An Ethnography of a Modern Job*. Edited by Stephen R. Barley. Ithaca, NY: Cornell University Press.

Pentland, Brian T. 1997. "Bleeding Edge Epistemology: Practical Problem Solving in Software Support Hot Lines." In *Between Craft and Science: Technical Work in U.S. Settings*, ed. Stephen R. Barley and Julian E. Orr, 113–128. Ithaca, NY: Cornell University Press.

Perrow, Charles. 1999. *Normal Accidents: Living with High-Risk Technologies*. Princeton, NJ: Princeton University Press.

Perry, D., and R. DeMaria. 2009. *David Perry on Game Design: A Brainstorming Toolbox*. Boston, MA: Course Technology.

Pickering, Andrew. 1995. *The Mangle of Practice: Time, Agency, and Science*. Chicago: University of Chicago Press.

Pinch, Trevor, and Wiebe Bijker. 1989. "The Social Construction of Facts and Artifacts: Or How the Sociology of Science and the Sociology of Technology Might Benefit Each Other." In *The Social Construction of Technological Systems: New Directions in the Sociology and History of Technology*, edited by Wiebe Bijker, Thomas P. Hughes, and Trevor Pinch, 17–50. Cambridge, MA: MIT Press.

Pinch, Trevor, Harry M. Collins, and Larry Carbone. 1997. "Cutting Up Skills: Estimating Difficulty as an Element of Surgical and Other Abilities." In *Between Craft and Science: Technical Work in U.S. Settings*, edited by Stephen R. Barley and Julian E. Orr, 101–112. Ithaca, NY: Cornell University Press.

Postigo, Hector. 2003. From Pong to Planet Quake: Post-Industrial Transitions from Leisure to Work. *Information Communication and Society* 6 (4): 593–607.

Postigo, Hector. 2007. Of Mods and Modders: Chasing Down the Value of Fan-Based Digital Game Modifications. *Games and Culture* 2 (4): 300–313.

Pritchard, Matt. 2003. "Ensemble's Age of Empires." In *Postmortems from Game Developer*, edited by Austin Grossman, 63–74. New York: CMP Books.

Radd, David. 2007. "ESA Applauds San Diego Raid." Edited by Libe Goad. GameDaily.Biz. Accessed July 20, 2007. http://biz.gamedaily.com/industry/news/?id=16634.

Ragaini, Toby. 2003. "Turbine's Asheron's Call." In *Postmortems from Game Developer*, edited by Austin Grossman, 299–309. New York: CMP Books.

Rheinberger, Hans-Jörg. 1997. *Toward a History of Epistemic Things: Synthesizing Proteins in the Test Tube*. Stanford, CA: Stanford University Press.

Rogers, Everett M., and Judith K. Larsen. 1984. *Silicon Valley Fever: Growth of High-Technology Culture*. New York: Basic Books.

Ross, Andrew. 2003. *No-Collar: The Humane Workplace and Its Hidden Costs*. Jackson, TN: Basic Books.

Sakamoto, Yoshio, Norikatsu Furuta, Kenji Imai, Hironobu Suzuki, Makoto Katayama, Koichi Kishi, Yumiko Morisada, and Hiroshi Tanigawa. 2000. U.S. Patent No. 6601851. Washington, DC: U.S. Patent and Trademark Office. Assignee: Nintendo Co., Ltd.

Salen, Katie, and Eric Zimmerman. 2004. *Rules of Play: Game Design Fundamentals*. Cambridge, MA: MIT Press.

Scarselletta, Mario. 1997. "The Infamous 'Lab Error': Education, Skill, and Quality in medical Technicians' Work." In *Between Craft and Science: Technical Work in U.S. Settings*, edited by Stephen R. Barley and Julian E. Orr, 187–209. Ithaca, NY: Cornell University Press.

Schadt, Toby. 2007. Postmortem: Not Your Typical Grind—Tony Hawk's Downhill Jam for Wii." *Game Developer Magazine* 14 (1): 30–37.

Schaefer, Erich. 2003. "Blizzard Entertainment's Diablo II." In *Postmortems from Game Developer*, edited by Austin Grossman, 79–90. New York: CMP Books.

Schell, J. 2008. *The Art of Game Design: A Book of Lenses*. New York: Elsevier.

Schüll, Natasha Dow. 2005. "Digital Gambling: The Coincidence of Desire and Design." *Annals of the American Academy of Political and Social Science* 597 (1): 65–81.

Schüll, Natasha Dow. 2012. *Addiction by Design: Machine Gambling in Las Vegas*. Princeton, NJ: Princeton University Press.

Scott, James C. 1976. *Moral Economy of the Peasant: Rebellion and Subsistence in Southeast Asia*. New Haven, CT: Yale University Press.

Sheff, David. 1993. *Game Over: How Nintendo Zapped an American Industry, Captured Your Dollars, and Enslaved Your Children*. New York: Random House Inc.

Sicart, Miguel. 2008. "Defining Game Mechanics." *Game Studies* 8 (2). Accessed December 15, 2008. http://gamestudies.org/0802/articles/sicart.

Simon, Roger. 2001. *Gramsci's Political Thought: An Introduction*. London: ElecBook.

Smith, Dorothy E. 1999. *Writing the Social: Critique, Theory, and Investigations*. Buffalo, NY: University of Toronto Press.

Smith, Marcus. 2007. "Postmortem: Resistance: Fall of Man." *Game Developer Magazine* 14 (2): 28–36.

Smith, Vicki. 2001. "Ethnographies of Work and the Work of Ethnographers." In *Handbook of Ethnography*, edited by Paul Atkinson, Amanda Coffey, Sara

Delamont, John Lofland, and Lyn H. Lofland, 220–233. Thousand Oaks, CA: Sage Publications.

Snow, Blake. 2007. "The Future of Video Games According to 1982." Joystiq. Accessed January 30, 2007. http://www.joystiq.com/2007/01/30/the-future-of-video-games-according-to-1982/.

Sony Computer Entertainment, Inc. v. Lik Sang International, LTD. 2003. HCA 3583/2002.

Sprigman, Chris. 2002. "The Mouse That Ate the Public Domain: Disney, the Copyright Term Extension Act, and *Eldred v. Ashcroft*." FindLaw. Accessed April 25, 2007. http://writ.news.findlaw.com/commentary/20020305_sprigman.html.

Staff. 2007. "2006 Game Developer Salary Survey Reveals Industry Trends." Gamasutra. Accessed July 29, 2007. http://www.gamasutra.com/php-bin/news_index.php?story=13352.

Staff. 2009. "Game Developer Reveals Top 20 Publishers, Debuts 2009 Research." Gamasutra. Accessed March 23, 2011. http://www.gamasutra.com/php-bin/news_index.php?story=25506.

Staff. 2011. "Worldwide Yearly Chart Index." VGChartz. Accessed May 21, 2014. http://www.vgchartz.com/yearly/.

Suchman, Lucy. 1995. Making Work Visible. *Communications of the ACM* 38 (9): 56–64.

Suchman, Lucy, Jeanette Blomberg, Julian E. Orr, and Randall Trigg. 1999. "Reconstructing Technologies as Social Practice." *American Behavioral Scientist* 43 (3): 392–408.

Sutton-Smith, Brian. 1998. *The Ambiguity of Play*. Cambridge, MA: Harvard University Press.

Taussig, M. 1999. *Defacement: Public Secrecy and the Labor of the Negative*. Stanford, CA: Stanford University Press.

Taylor, T. L. 2006a. *Play between Worlds: Exploring Online Gaming Culture*. Cambridge, MA: MIT Press.

Taylor, T. L. 2006b. "Does WoW Change Everything?: How a PvP Server, Multinational Player Base, and Surveillance Mod Scene Caused Me Pause." *Games and Culture* 1 (4): 318–337.

Train, Tim. 2003. "Postmortem: Big Huge Games' Rise of Nations." *Game Developer Magazine* 10 (7): 36–41.

Traweek, Sharon. 1988. *Beamtimes and Lifetimes: The World of High Energy Physicists*. Cambridge, MA: Harvard University Press.

Traweek, Sharon. 2000. "Fault Lines." In *Doing Science + Culture: How Cultural and Interdisciplinary Studies are Changing the Way We Look at Science and Medicine*, edited by Roddey Reid and Sharon Traweek, 21–48. New York: Routledge Press.

Tsing, Anna Lowenhaupt. 2005. *Friction: An Ethnography of Global Connection*. Princeton, NJ: Princeton University Press.

Turkle, Sherry. 1997. *Life on the Screen: Identity in the Age of the Internet*. New York: Simon & Schuster.

Turnbull, David. 2000. *Masons, Tricksters, and Cartographers: Comparative Studies in the Sociology of Scientific and Indigenous Knowledge*. New York: Routledge.

Varma, Roli. 2006. *Harbingers of Global Change: India's Techno-Immigrants in the United States*. Lanham, MD: Lexington Books.

Wark, McKenzie. 2007. *Gamer Theory*. Cambridge, MA: Harvard University Press.

Waters, Darren. 2007. "EA Wants 'Open Gaming Platform.'" BBC News. Accessed October 23, 2007. http://news.bbc.co.uk/2/hi/technology/7052420.stm.

Weber, Tim. 2007. "Games Industry Enters a New Level." BBC News. Accessed June 4, 2007. http://news.bbc.co.uk/2/hi/business/6523565.stm.

Wen, Howard. 2007. "Analyze This: Will There Ever Be One Console to Rule Them All?" Gamasutra. Accessed November 16, 2007. http://www.gamasutra.com/view/feature/2012/analyze_this_will_there_ever_be_.php.

Whalley, Peter, and Stephen R. Barley. 1997. "Technical Work in the Division of Labor: Stalking the Wily Anomaly." In *Between Craft and Science: Technical Work in U.S. Settings*, edited by Stephen R. Barley and Julian E. Orr, 23–52. Ithaca, NY: Cornell University Press.

Whitehead, N. L., and R. Wright. 2004. *In Darkness and Secrecy: The Anthropology of Assault Sorcery and Witchcraft in Amazonia*. Durham, NC: Duke University Press.

Williams, Dmitri. 2002. "Structure and Competition in the U.S. Home Video Game Industry." *International Journal on Media Management* 4 (1): 41–54.

Wilson, Elizabeth A. 1998. *Neural Geographies: Feminism and the Microstructure of Cognition*. New York: Routledge Press.

Wilson, Phil. 2007. "Postmortem: Realtime Worlds' Crackdown." *Game Developer Magazine* 14 (9): 26–31.

Wilson, Trevor. 2008. "Game Developer's Magazine's Top 20 Publishers for 2008." Gamasutra. Accessed March 23, 2011. http://www.gamasutra.com/view/feature/3800/game_developer_magazines_top_20_.ph.

Wong, Nicole. 2006. "Exclusive: Nicole Wong Reveals Identity of EA Spouse." The Mercury News. Accessed April 25, 2006. http://web.archive.org/web/20060430102757/http://blogs.mercurynews.com/aei/2006/04/exclusive_nicol.html.

Wyatt, P. 2012. "StarCraft Was Once a Busted Pile of Crap." Accessed November 11, 2012. http://kotaku.com/5942128/starcraft-was-once-a-busted-pile-of-crap.

Zabusky, Stacia E. 1997. "Computers, Clients, and Expertise: Negotiating Technical Identities in a Nontechnical World." In *Between Craft and Science: Technical Work in U.S. Settings*, edited by Stephen R. Barley and Julian E. Orr, 129–153. Ithaca, NY: Cornell University Press.

Zichermann, Gabe. 2011. "Gamification Is Here to Stay." The Atlantic. Accessed August 29, 2011. http://www.theatlantic.com/technology/archive/2011/08/gamification-is-here-to-stay/244232.

Index

The letter *f* following a page number denotes a figure, *b* denotes a term found in a box, and *t* denotes a table.